ONCE UPON A RAVEN'S NEST

Also by Catrina Davies

Homesick: Why I Live in a Shed
Fearless

ONCE UPON A RAVEN'S NEST

Catrina Davies

riverrun

First published in Great Britain in 2023 by

riverrun
An imprint of

Quercus Editions Ltd
Carmelite House
50 Victoria Embankment
London EC4Y 0DZ

An Hachette UK company

Copyright © 2023 Catrina Davies

The moral right of Catrina Davies to be
identified as the author of this work has been
asserted in accordance with the
Copyright, Designs and Patents Act, 1988.

The publisher would like to thank Hannah Scrase for permission to use
the recordings and transcripts of the author's interviews with Hedley Ralph Collard
undertaken on various dates between 2014 and 2020.

The recordings and transcripts of Catrina Davies' interviews with Hedley Ralph Collard
between 2014 and 2020 © Hannah Scrase, 2023.

All rights reserved. No part of this publication
may be reproduced or transmitted in any form
or by any means, electronic or mechanical,
including photocopy, recording, or any
information storage and retrieval system,
without permission in writing from the publisher.

A CIP catalogue record for this book is available
from the British Library

Hardback ISBN 978 1 52942 499 7
Ebook ISBN 978 1 52942 500 0

Every effort has been made to contact copyright holders.
However, the publishers will be glad to rectify in future editions
any inadvertent omissions brought to their attention.

Quercus Editions Ltd hereby exclude all liability to the extent permitted by law
for any errors or omissions in this book and for any loss, damage or expense
(whether direct or indirect) suffered by a third party relying on
any information contained in this book.

10 9 8 7 6 5 4 3 2 1

Typeset by CC Book Production
Printed and bound in Great Britain by Clays Ltd, Elcograf S.p.A.

Papers used by riverrun are from well-managed forests and other responsible sources.

For Hannah

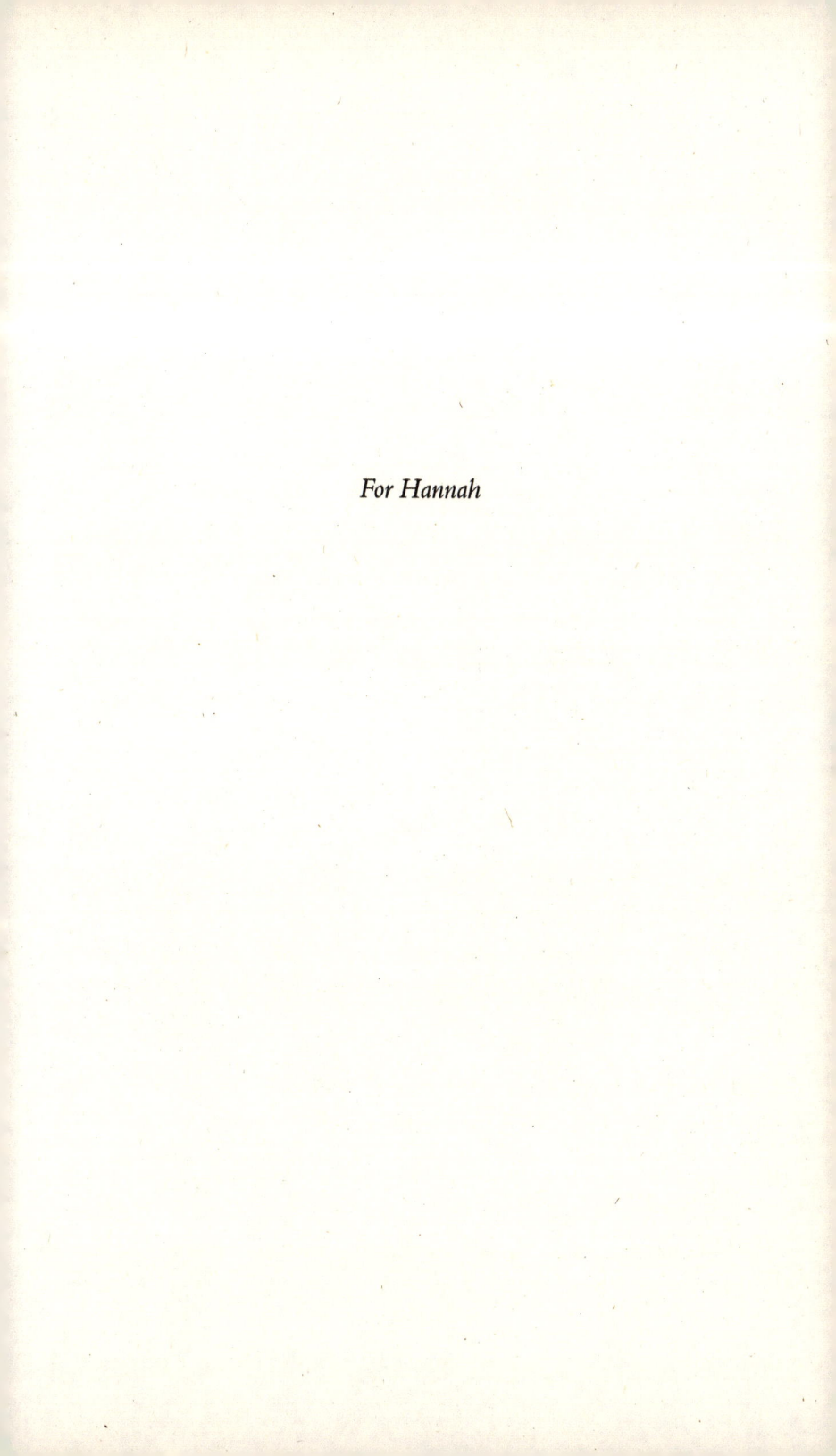

Preface

Once Upon a Raven's Nest is based on the recollections of Hedley Ralph Collard, which I recorded in Wales and on Exmoor between 2014 and 2020. While his recollections form the backbone of this story, I have altered, edited and added to the recordings, both to protect the identities of people involved, and to make them work as a coherent narrative. At his request, Ralph has become Thomas Arthur Hedley, and where necessary I have put my own words into Thomas Hedley's mouth, and filled gaps in his story using my own imagination. This is a portrait, not a biography.

I met Ralph in 2014, when I was house-sitting in Wales, near where he lived. We became friends, and the following year I stayed with him for ten days while his partner was away. I continued to stay with him for a few weeks every year until 2020.

The more time I spent with Ralph, the more fascinated I became with his life. It wasn't just his accident and its aftermath, but the decades leading up to it. His childhood on Exmoor, memories that spoke of the rapid changes that occurred in rural England during the latter half of the twentieth century.

His deep love for the women and places that defined him. His individuality, which often collided with his need to belong. The tension between his attraction to machinery and his connection to nature. His love of animals – and his desire to hunt them.

I was haunted by the feeling that the particular circumstances of his life somehow expressed something universal, and urgent, about all of our lives at this moment in history. The way we hurtle around, pushing and pushing at the boundaries, the limits of what's possible, until we pass the point of no return, and everything we take for granted is lost. Yet still we have to keep on living.

With his permission, I began recording Ralph's stories. Then, also with his permission, I began trying to turn them into a book.

It's one thing to imagine a book, quite another to overcome all the technical obstacles that stand in the way of it becoming real, not to mention the ethical ones.

Reality is made of randomness, one damn thing after another, with no beginning or end, and no inherent meaning. Books and lives, on the other hand, do have beginnings and endings, and within the limits of their forms, meaning can be constructed. The problem of whether this constructed reality is *true* is probably best left to quantum physics. In the same way that the truth of particulate behaviour changes depending on who is looking, the truth of stories depends on who is telling them, and who is listening.

You could argue that even the made-up parts of this book are true, in so far as they were dredged up from my own lived experience, enriched by the time I spent with my friend. Equally, you could argue that even the true parts of this book are made up, in so far as they are based on memories, smoothed by time and expressed in language, which can only ever approximate.

Beyond argument is the fact that I owe a huge debt of gratitude to Ralph, for his remarkable voice, poetic turns of phrase, deep tree-knowledge, friendship, and support.

I trust he knew that the making of this book was an act of love.

Most of us walk unseeing through the world, unaware alike of its beauties, its wonders, and the strange and sometimes terrible intensity of the lives that are being lived about us.

Rachel Carson, *Silent Spring*[1]

4.5 Billion Years Ago

A cloud of dust and gas spins in the void, and a sun forms in the centre of the cloud, and the remaining material – the small particles of swirling dust and gas – draw together, and are bound by gravity into larger particles, and solar winds sweep away the lighter elements, like hydrogen and helium, leaving only heavy rocks, and small worlds are formed out of these spinning rocks.

3.7 Billion Years Ago

Life starts on Earth: first microscopic organisms, then bacteria, then corals and shelled brachiopods, who come to fill Earth's shallow waters.

440 Million Years Ago

Ice advances over Earth's southern supercontinent. Sea levels fall, habitats are broken, food is scarce. 85 per cent of life is lost.

365 Million Years Ago

An armoured fish, 33 feet long, with a helmet of bone-plates and a fang-like cup on its jaw, patrols Earth's teeming waters. On the dry edges of the oceans plants evolve, and they make roots, and their roots transform the land from rock and rubble to nutrient-rich soil, and the soil runs into the oceans, and causes algae to bloom on an enormous scale, and the armoured fish suffocates, along with all its prey. 75 per cent of life is lost.

250 Million Years Ago

Explosive volcanoes release carbon dioxide into Earth's atmosphere, causing a greenhouse effect. Weather patterns shift, sea levels rise, acid rain beats down upon the land. 90 per cent of life is lost.

210 Million Years Ago

Volcanoes erupt again. Carbon dioxide heats the atmosphere. Permafrost melts and trapped methane is released. Ice melts, sea levels rise, oceans acidify. Giant crocodiles perish. Early dinosaurs perish. 85 per cent of life is lost.

65 Million Years Ago

An 8-mile-wide asteroid crash-lands at a speed of 45,000 miles per hour, punching a hole 110 miles wide and 12 miles deep in the surface of the Earth, in the place we now call the Yucatán Peninsula, in Mexico. Debris and dust is hurled into the atmosphere, turning the sky black. Plants wither and die. Dinosaurs, who have been ruling the Earth for 180 million years, are wiped out. 75 per cent of life is lost.

12,000 Years Ago

Earth starts to warm. The glaciers of the late Palaeolithic retreat. Tundra gives way to forest. Mammoth and woolly rhinoceros die. The Holocene starts.

10,000 Years Ago

Humans in the Fertile Crescent[2] start to cultivate wheat and barley, and their techniques spread throughout the Indo-European world. Humans in Central and South America grow maize, bottle gourds, squash and beans. Sheep and cattle are domesticated. The human population begins to increase.

1712

Thomas Newcomen, born an ironmonger, called to the pulpit as a Baptist lay preacher, invents the first practical fuel-burning engine. Unlike early devices powered exclusively by water, these steam engines are powered by coal.

1945

The first nuclear weapon is detonated in the Jornada del Muerto desert, about 35 miles south-east of Socorro, New Mexico.

1953

A heavy storm in the North Sea damages 900 miles of UK coastline and kills 307 people. Sea walls are breached in 1200 places, and 160,000 acres of land are underwater. 200 people are forced to abandon their homes in London, and take refuge in Canning Town Public Hall.

Thomas Arthur Hedley is born in Taunton hospital.

15 March 2009

I lie motionless for twenty minutes, waiting for the ambulance to arrive. Two paramedics. Take one look at me and send for another ambulance. Cover me in a blanket made of silver foil.

My mind swims down and backwards, finds a gate leading into a patch of woodland. The gate is rusty, sunk on its hinges. I have to lift it slightly to open it. I hear the sound of the latch going and the sound of water and the sound of bees. I hear the sound of trees rustling and swaying in the wind. I hear Hope's voice, telling me to stay with her.

Second ambulance takes forty minutes to arrive.

I listen for squirrels and small birds rummaging for food. I listen for rooks and crows and woodpeckers. I listen for all the sounds I've heard throughout my life, and I listen for new sounds, things I never acknowledged at the time, because I never really knew they were there.

Four of them carry me out on a stretcher. I'm trying to say goodbye to the dogs. Spike licks my hand, ever so gentle. I hear one of the paramedics saying do we need bells, and another one saying no.

My mind takes me through the patch of woodland and out into an area of bracken. I find a narrow footpath, like a sheep's path, winding up a steep slope.

The hospital is loud and bright. Nurses come every four hours to lift and turn me. My mind is a jungle, flitting from here to there, dipping in and out of time. One big mass of memories, seasons all crushed together.

I keep my eyes closed and list everything I see. Swallows, willow-herb, heather, gorse. I stop to ask a slow-worm if he's got a mate anywhere. No, he says. The grass is terrible green, but the sky is clearest blue like it has never heard of rain. I follow the path as it winds its way upwards, through the woodland and the bracken and out into a meadow. I use my mind to keep on climbing.

Sometimes a nurse comes in and starts poking around before I get to the top, or another patient starts shouting and screaming and I lose my way. I have to go back and retrace my steps.

I thought I'd be dead before I turned thirty. Bang, lights out. I couldn't imagine being myself, Thomas Arthur Hedley – *young Thomas* – but old. It was easier to imagine being dead.

There are flowers on the high ground. I close my eyes and open the gate, and I can see the whole of my life spread out in front of me, like the opening shot of a film about some ancient war.

I can see my grandfather on Father's side, waving his shotgun at the contractors who bulldozed his farm. Fisher's Mede, Great Meadow, Bath Meadow. I can see the farmhouse where Father grew from a boy into a man, all ancient and ruined and sagging in on itself. I can see the secondary modern, my own school, which they built on the field where Grandfather used to keep the fat stock.

I can see the village of Winsford, where Mother was borned, and Mother's mother, and her mother's mother, going back hundreds of years. I can see my grandfather on Mother's side, climbing out of the Welsh valleys covered in coal, working his way west, part of the gang that was building the roads.

I can see Mother and Father courting. Mother going round and round the fields she knew like the back of her hand, pixie-led, or so she'd have it. Father walking drunk along the road to Dulverton, flinching at the sound of a sheepdog's footsteps, fleeing his own shadow.

I can see black dogs floating down the Exe valley, grandmothers turning into hares and back again, running over the moor with packs of hare hounds chasing them. I can see hedges full of foxgloves the foxes used to stalk their prey, swallows plucking celandine to mend their blindness.

I can see the Devil sunbathing on Tarr Steps, getting into

a cussing match with the vicar, wilting the trees with his language.

And every time I open my eyes Hope is there.

Stay with me, she says.

I work to bring up the exact feeling of being cold, warming. I can't do it. Legs like tombstones. I can't tell if they're frostbitten or scalding hot. I can't feel the presence of a fly, or twitch it away.

All I can do is hurt, and I can't move to ease it.

I close my eyes and climb, up and up and up, over and over and over, through the hunting gate, through the patch of woodland, through the area of bracken, away from the nurses and lights and machines. I learn to focus on every detail. It's a way to control my mind, keep from drowning.

And sometimes, when I get high enough, I can see how life is less of a sheep's path winding itself up the side of a hill from the bottom to the top, and more like the patterns clouds make on the surface of the sea. I can see Gracie and Gillian, Johnny and Alfie, Uncle Jim. I can see foxes running, stags roaring, gorse flowers opening. It's all happening at once.

I go up and onto the high ground, and it's like a fog has lifted and I can see how everything is made up of a long chain of events that might easily not have happened. One thing leading to the next thing, and the next thing, and on and on and on. Small things most of them, forgotten as soon as they are done.

It's only when you add up all these small things that you get a sum so fraught with heartache you can't even start to reckon on how much you've lost.

23 September 2009

Scientists from the Stockholm Resilience Centre publish a paper setting out nine planetary boundaries, or limits, beyond which we can't push Earth's systems without risking the collapse of human civilization.[3]

Rate of biodiversity loss, land-use change, aerosol pollution, novel entities, ozone-layer damage, ocean acidification, nitrogen and phosphorus cycles, freshwater cycles, and climate.

They argue that three of these boundaries have already been crossed: biodiversity, nitrogen/phosphorus, and climate, with the remaining six in a critical condition.

The breaches have all occurred during Thomas Hedley's lifetime, a period of untrammelled economic growth known as the Great Acceleration.

PART ONE
BREAKING BOUNDARIES

TYBALT: What wouldst thou have with me?
MERCUTIO: Good king of cats, nothing but one of your nine lives.

>William Shakespeare, *Romeo and Juliet*

2014

I wouldn't be able to cope.

Catrina's on the swing I hung off the big beech in front of the cottage. Had to put my harness on just to get up to the lowest branch. She's twirling one way then the other, tangling the ropes this way and that way, hither and thither. She's one of them as can't sit still.

You would, I tell her. We cope with what we have to.

She's writing a book, something about living in a galvanized shed and how nobody got houses no more.

I got to bite my tongue. I worked my fucking bollocks off for my little cottage on the outskirts of Dulverton, and that was after my heart attack.

If you work hard it'll pay off, I tell her.

No, she says, hard work don't count no more.

She'm argumentative like that.

Well, I says. I hear everyone wincing and whining about how they hant got this and they hant got that and not willing to do a decent day's work.

She says it's political.

Well, I says, I'm not interested in politics, except what bugger's going to come and kill me.

Hope asked her to come and light the fire and give me a bit of company so I'm not on my own too long. Problem is she never stops asking questions.

What's it like?

She drags her feet on the ground to stop the swing moving.

Well, I says. It's like a massive, massive punishment.

You know what the old saying is, you reap what you sow. Well I reckon I'm being repaid for some shit I done years ago, like punching Jack Thomas in the earhole, or running over his cat.

It's late afternoon and crows are raucous, they know something's coming.

Do you really think that?

Well, I says, I'd like to think life's got some fairness buried in it somewhere, do you know?

I lean my torso forwards and backwards again, but it don't make no difference.

The first raindrops fall and catch and break and splatter on the ground.

We better go inside, I says, and she gets up off the swing and follows me into the house. We go up the track and in through the back doors because I'll only use the lift in the evenings when it's dark.

She kneels on the floorboards by the wood burner and I offer up instructions on which pieces of wood to put in first and how much to open the vent.

Put the kettle on top, I says.

I tip the chair back and sit by the fire with my legs up until the kettle starts whistling.

I already taught her to pour my tea into the mug with the big handle. I can wedge my hands in through and lift him up like that.

I tip the chair up and roll over to the windows and look out across the valley at the rain.

I still can't hardly get a grip on it. I sit here sometimes by the window and I sort of like leave my body and hover up above, and it's like looking down on a stranger. It's me in the wheelbarrow and it's not me, like someone has taken me out of my box and put me in a new one. Like a dog suddenly turned into a cat, like the women on Exmoor who turned into hares and back again, only this time there's no turning back.

Catrina comes and sits on the sofa. It's the one as used to be in the Red Lion in Dulverton, the one Winston Churchill sat on.

She's got her notebook with her and a pen.

The reason I think it could be fate, I says, is I've been angling for it since the day I was borned. Weighed five pound and fell down to three. Gave the doctors and nurses a rake of trouble and I been ought but trouble ever since. I had accidents and accidents and accidents. Broke my thigh, broke my arm, broke my ankles, both of them, broke my wrist, my thumb, two fingers, a couple of ribs. Big Steve called me up in the hospital.

You really bloody done it this time, he says, and I says yeah, and I think shit what more.

Life to me meant wanting to earn enough money so I wasn't scraping the barrel. I wanted a family, somebody to protect. I wanted to be left alone by all the bureaucracy and shit. I wanted to enjoy myself. I wanted to live.

The rain is hammering on the windowpanes. I'm worried for Hope driving all that way back over the hills.

I spin the chair around.

I crashed a car once between Exford and Lucker Bridge. It was a Triumph Herald, and they had a roof you could just unbolt and take off. I tipped him over and pulled the roof off. It was just the backs of the seats left in. He kept going really fast, and I was down in the footwell.

You were upside down?

Well, I says. Somat like that.

Catrina's scribbling in her notebook.

I got to stop talking, I says. I'm getting wored out.

Only I don't stop.

We carry on like that for the rest of the winter and the ones that come after. Me telling up my tales and her writing them down in her notebook, scribbling like the world's going to end and she'm got to get it all down first.

And sometimes with the telling of them the tales start to reach for something I can't wholly fathom, and it's like being up on that hill again and I can see the whole of my life spread out in front of me, clear as light.

And other times the words are like the scattered bones of an animal that lived and died a long, long time ago, and all around him it's dark as darkest night.

9

The world has lost two-thirds of its wildlife in the past fifty years.[4]

1953

I'm hatched in Taunton hospital, five weeks early and so small they store me in a jam jar like a pickled onion. Incubator. Parents take me home to Exmoor, which is where they come from. They say Devil got thirsty, and he dug for water, and he made Dunkery with a spadeful of earth, and he drank the water out of the hole he left, and they called the hole Winsford, and that's Mother's village.

There was all of them borned in the one cottage. Betty, Olive, Mother, Doris, Arthur, Fred, Mervin, Marion, Gwen, Dor, Sybil, Else – only nobody knows if Else was a cousin or an aunt, six times removed or just the once. She'd dance on the bridge when the river was in spate, set fire to herself, eyes like flying saucers.

It's a tea room now, the cottage where Mother was borned. After Father got ill it was my job to go and cut the grass and put out the rubbish. That's what we kids done to earn a bit of extra pocket money. Half a crown we got to mow the lawn, and it took us half a Saturday.

Father's father's tough as custard and twice as yellow. He can pick up a granite roller and load it into a butt cart. Got a farm near Dulverton. Likes to walk into the pub with his sleeves rolled up, itching for a fight. Then the council kills him. Compulsory purchase. Take his land.

He goes down waving his shotgun at the contractors, only they say it isn't his land no more. He keeps the farmhouse, but they build new houses all around, council houses and roads, and that's all the good ground gone.

Father's brother Jim is my favourite. His son Keith is the brainy one, but I'm the wild one. Rufty-tufty, and I'm into hunting, and him a gamekeeper. Jim's a yes-sir man, because keepers got to doff their hats at shooters, only that goes against my grain. I think we ought to be slightly more equal than to go around doffing our hats.

Jim works on the Wingfield Estate near Trowbridge. It's a hell of a way to go with the reliability of cars, but Father likes his driving. He'll sit behind the wheel for a week, think nothing of it. Golden times, especially in the light of what comes after.

Mother's mother's a Gardner, from Howtown, old granny Gardner, born in the late nineteenth century. There've been Gardners on Exmoor since the 1700s, so we're going back a bit.

Place is crawling with aunts and uncles. They smoke the bloody house out and tell up their tales, and I sit up and listen, quiet as a mouse, until I'm scared to go to bed. Bloody shit. Or I'm flicking through *Gulliver's Travels* and having nightmares about that, because all these little men are stealing my bedclothes and tying me down.

1957

I'm four years old when a chap called Mr Fox comes calling. There's a knock on the door and Mother says Father look, it's Mr Fox and then she says to me and my sister, you chillers go upstairs and don't you come down again until I say so.

Well I don't want to go upstairs, not while I'm thinking there's a fox in the bloody kitchen, so I hide behind the curtain in the living room and I hear Mr Fox come in and sit down and talk a bit with Mother and Father and he sounds like all the other adults do and I can't hardly believe he'm a fox at all.

I got to see with my own eyes so I creep out from behind the curtain, only I trip and fall over and Mother comes running out and that's when she puts me in the cupboard under the stairs and bugger me there's spiders in there bigger'n my hand. I'm terrified in case they'm going to eat me whole.

Father lets me out and I can't even cry.

You shouldn't have put en in there, he says to Mother.

Well, she says, twill teach en a lesson.

I'm a strange boy, everyone says it. I like nothing better than to sit under a tree and close my eyes and listen and wait for birds to come and sit on my hand. Robins and blackbirds and sparrers. What are you bloody staring at, Mother'll holler, and I'll be stuck for words to describe the feeling I get from looking at the bare branches of a hawthorn in December.

The landlord up at Molden Inn's got wild animals in pens. He's got Fox, and Badger, and Owl, and Parrot, and Monkey. He's got a red stag called Ernest, and one day when there's lots of people around he lets him out. Stag makes a beeline for me, puts his hooves on my shoulders and knocks me over.

When I go to bed that night I see him in the wallpaper, and I'm screeching and screaming, and Mother comes running up.

Turn over, she says, and look at something else.

But he'm everywhere, I says, and he'm not friendly.

1958

Scientists start measuring atmospheric CO2 at the Mauna Loa mountaintop observatory in Hawaii. The CO2 level stands at 316 parts per million (ppm), which is slightly higher than the pre-industrial level of 280 ppm.[5]

1959

It's coming up to Christmas and I'm woken by the clattering of a lorry in the dead of night. I creep over to my bedroom window and look out and there's a crowd of sheep in the lane and Father's shouting and there's one of my uncles shouting back.

I find out later he'd been out rustling only he loaded up the sheep in such a hurry he didn't do the tailgate up proper, and right outside the cottages the tailgate comes open and all the sheep jump out. Father he'm furious. Don't see much of Uncle after that.

1960

I'm seven when the real trouble starts. I pull Millie Steer's knickers down in the school playground. Disgusting boy.

It's her friend goes and tells on us. Debbie Simcock. I catch up with her years after. You bugger, I says, you cost me a suspension and twelve months of lunch breaks. I had to do my ten times tables, right up to twelve times. Forgot them all now.

They send me home with Mother, only she don't want me, so I traipse up to Knaplock with Father, and there's plenty more trouble waiting for me up there.

Father says if Granny offers you a cup of tea or anything to eat, you say no. I got your sandwiches here, he says. Well we go in for lunch and Granny's got the table all laid out and she'm offering me tea and cake, and I got to say no, and I'm hungry. I'm always hungry, I'm a scrawny little thing. I'm angry with Father, but I think I know what he's on. There are chickens running around and pecking at the bread, and a cat with a runny nose drinking out of the milk jug, and a large white sow fast asleep under the table.

Have you ever seen a full-growed large white sow? Monster.

I rest my feet on the sow until Derek the farmer comes in. He sits on a kind of swivelling secretary seat at the head of the table and he kicks and kicks to get the pig out from underneath. Well as her stands up she nearly picks up the table and carts en off.

Derek's one of Exmoor's wildest. Gurt raw-boned bugger with a flat cap, as broad as he's tall, don't care for nobody. Neighbouring farm's got a stallion, and one day Derek comes into the field and the stallion comes after him and Derek gets off his horse and stands his ground, and when the stallion gets right to him, Derek punches him on the end of his nose, and the stallion falls over.

He's got a pack of hounds for hunting and he's got them half starved in a cage to keep them hungry. I've got a stick and I like to rattle it up against their cage to piss them off, only one day they've had enough and they surge towards the galvanized door and break it and take off after me. The next thing Father sees is me running up the flatbed and jumping into the lorry and the hounds all clustered round baying for my blood. Derek shoots them after that, every last one of them.

He's got a County Crawler. It's based on a Fordson Major, only with tracks instead of wheels. Him and Father go blasting stone down by the river, drive down over the cleave with the Crawler in the trailer, throw a load of explosives, then run and hide. It's my job to go around picking up loose stones.

Then one day Father says no.

You can't come today, he says, it's too dangerous.

Well I kick up hell, but they won't budge, so I wait and follow them down on my own. I sneak down and hide in the woods, and after the blasting stops, I go over to inspect the damage, only there is no damage. I go along until I get to the pool. There's a pool down there, a very deep pool, about fifty metres deep, and long, and all along the edges are dead fish, piles of them.

It's not until years later I'm told there was a load of salmon in the pool and Derek and Father had gone down and chucked explosives in, to get the salmon out.

Next day the water bailiffs come down to the wood opposite, only Derek fires into en with a .22 rifle and that scares them away. That's the kind of tough-assed bugger he is. I'm in awe of him right up until he gets wored out and has a heart attack.

I wear Harris tweed shorts and my knees get so cold they split and bleed. Father has to rub cow salve on them and carry me upstairs to bed, but I don't care. I have some lovely times up there on that farm, just mooching around. Fields back onto open moorland. Father'll say don't you wander too far, then forget all about me until lunchtime.

One day I ketch a wood pigeon with my bare hands. I look up into a tree and there's a wood pigeon just sitting there. I creep up and creep up and he never flies away. I snatch him off the branch and cart him up to Father, chuffed as nuts. I'm stroking him and kissing his head, and Father he'm shouting.

Put it on the ground and don't touch your face.

But he's my little friend, I says.

And Father answers by picking up a stick and knocking the pigeon's head off. Tisn't very friendly. He makes me wash my hands in the stream until there's nought left on en, not even skin. Then he takes the stick and points to a dead fox in the grass, then he points upwards and there's bloody birds dropping out of the sky.

Strychnine, he says. Crows came in and stripped the fields and Derek soaked the verges in revenge on the crows.

Mother she'm furious with Father.

Strychnine kills nine times, she says, and even cats only got nine lives.

I got to stay home with Mother after that, and that's why I hide in a raven's nest.

1960

I want to be out the front in the lane riding the pedal car Uncle Edowin has given me, but the tyres have worn out and Mother can't abide the noise of the bare metal wheels scraping the tarmac, so I'm out the back playing with Billie Jean, Mother's goat, only he's tethered to a post so he don't eat the sunflowers and the cabbages and the peaks of new potatoes Father has planted. Goats are a menace, says Mother, and so are children, and she wishes she'd never been lumbered with either.

I feel sorry for Billie Jean not being able to reach the nice green grass so I trip backwards and forwards between the lawn and the post where he's tethered, ripping up handfuls of grass and feeding it to him. Good boy, I says, and he butts me in the chest, only he don't have much in the way of horns, and we'm friends. It isn't until the doorbell chimes that I notice what I've done to Father's lawn. Father's a farmer's son, through and through, and everything he tends is perfect.

I can hear Mother's footsteps going down the hall to open the door. It's the ironmonger. We got no shops or supermarkets, only delivery vans. There's Kipper the fisherman, whose real name is Ernie, like the stag up at Molden Inn. There's the butcher, who's got no fridge to store his products, so the meat is black by the time we come to eat it. The ironmonger is my favourite. He sells everything. Ajax, welly boots, dungarees, waterproofs, billhooks, pickaxes, forks, longstick shovels, bleach, clothes lines, clothes pegs, hot-water bottles, rope – sisal rope, not plastic shit – shoes. I've outgrown all my shoes, which is why I'm not wearing any.

Tommy! Mother is shouting.

She's holding up a pair of black plimsolls, only the ironmonger takes one look at my bare feet and pulls out a rag from the pocket in the front of his apron. I sit on the floor and wipe the mud off with the rag, then push my feet into the plimsolls. I tie the laces carefully, like I've been taught. Over, under, loop. Mother's impatient. She kneels on the floor and pokes and prods at my feet.

There, she says. They should last about as long as Harry Shaddock bide to London with a coal of fire up his ass.

I stand in the doorway after the ironmonger has driven off. I want to go with him, catch a ride to Porlock, make castles out of stones on the beach. I love the beach, although I've only seen it twice. Father's always working. Weekends, holidays, works every bloody day there is, evenings too. He's always out, always dirty. It's our job to tread on his overalls in the bath, keep them clean and get the ticks and lanolin out. It's what we do for entertainment.

I can hear Mother screeching, and the sound of her footsteps coming down the stairs, and the sound of the flippy stick smacking against her thigh, and that's when I remember what I've done to Father's lawn.

I'm out of the door and racing up the lane, racing past the elms on the corner, past the oak with the hole in the trunk where robin lives, past Darby's Knapp where the turtle doves live. I race up that lane as fast as my little legs will go, and Mother is chasing me as fast as her legs will go, and she's panting and mumbling and cracking that old ruler against her thigh.

I'm running along the side of the hedge looking for a place to dart inside and hide from Mother. It's a typical Exmoor hedge, banked up, with trees planted on top, been like it for hundreds of years. I'm looking all along the hedge for a place to hide and that's when I see the nest. It's like a basket made of twigs, a great pile of twigs in the uppermost branches of the tallest beech in the hedge.

Rook has a bald beak. Crow's completely feathered. Raven's much bigger and blacker and stronger built, and he honks. Raven's nest is deep, full of sheep's wool – mainly sheep's wool – and bailer cord.

I grab hold of the lowest branch and manage to pull myself up. Nest must have been fifty foot high, and I don't like heights, but I haul myself up, branch by branch. Getting in is the worst, because there isn't much to stand on, but I lever myself over and snuggle down with my legs hanging out the side and it's really cosy. I lie there rocking in the wind. Must have been an empty nest, or they'd have come back and bothered me.

I wake up to the sound of shouting.

Git down out of that tree.

I lean out over the edge of the nest and I can see Father down below. Bloody mass of trouble.

I'm not coming down, I says. You'll beat me.

Git down, says Father.

No, I won't.

Father counts to ten and then he turns and walks off down the lane. I can hear the pair of tawny owls who live at the end of the field behind the cottages calling to each other. The sun's gone only there's a big yellow moon.

Father comes back with an axe. He's got it over his shoulder. It's his big axe, the one he uses for oak and elm.

I'm going to ask one more time, he says, raising the axe so I can see the blade glinting in the moonlight.

No, I says. I'm not coming down, because you'll beat me.

Father stands there a moment or two longer, then he raises up the axe and he cuts the tree down.

8

Half of the world's forests have disappeared since 1950.[6]

2014

Twas about six months in when they told me I was going to be in pain for the rest of my life. Surgeon called it neuropathological. Pain without feeling. Signals come to where the breaks are, then they go everywhere, it's a dead end. I get pins and needles all down the backs of my legs and into my heels, stabbing and throbbing.

Catrina opens the door to the workshop and pulls out the ramp.
    We're looking for pieces of foam to stuff between the arms of the chair and my knees. I can tell the hard edges of the chair are digging into my skin and I don't want sores. Put me in bed for months. Seat needs adjusting.

First visit to the hospital was the time I broke my femur, I says, and I tell her to get hold of my legs.
    Pull em, I says. Pull em hard.
    Most people in my position don't get any pain, one in nine, something like that.
    She's got her hands around my legs above the knee.

Don't pull em off, I says, and she lets go and my legs flop back.

Not like that, I tell her. Now wedge the foam in. Try and get it right down.

I am trying.

You are.

What happened with your femur?

Christ, I says. Heard me screeching in bloody Tiverton.

1960

It's the first day back at school after the holidays and I'm wearing the new shirt Aunty Flo got me in Minehead the day Father was getting tested. I've got a new pair of short trousers too, only they'm so big they fall off every time I stand up. Mrs Evans made me tie them together with a piece of string, and that made Jimmy Carson howl. Mother says I'll grow into them, only I'm not very good at growing.

I've been wearing farm clothes all summer and I prefer them. Tattie sacks that keep my shoulders dry and the mud off. One over the shoulder, hooked in with a piece of wire, and one as a skirt, joined at the waist with another bit of wire. West of England sacks. Lord Carnarvon's liveries. Father wears them too, in the winter months, when the farm work goes slack and he spends his time hedging and digging trenches.

Father carried me home in a tattie sack once. I was helping tie the wool when they were shearing, only I jumped a gate and twisted my ankle and after that I couldn't walk. Father picked me up and put me in a tattie sack and slung it over his shoulder to carry me home. It was only later we discovered the sack was full of ticks. Mother had to dip me in a bathful of disinfectant.

I'd do anything to swap the new shirt Aunty Flo bought me for a couple of woven tattie sacks and a bathful of ticks. I'd even swap it for pulling mangels, which is not a job I like. It's my part to wind the handle and it's hard work, although I do like the sound the juicy mangels make when they go through the slicer. But mangels are for winter, and it's only September. Father'll be making hayricks and thatching them with rushes. I love making hayricks. I've got my own little scythe, and Father lets me cut rushes with it for the rick, and bundle them up.

I've not been in school since I pulled Millie Steer's knickers down. I was seven then, and now I'm eight, and I reckon I'm in a different category to the other children on account of all the farm work. They reckon it too, I expect, since none of them want to sit near me. Johnny sometimes sits with me, but Johnny's a farmer's son and never comes to school when it's harvest. I'm only a farm labourer's son. I begged Mother to let me keep going to work with Father, but she put her foot down. He's been out of school five months, she says. They'll send the authorities.

Winsford Primary School is more'n a hundred years old. There's a door for girls, and another one for boys, and two separate cloakrooms and one classroom with three different classes going on at the same time so it's hard to hear. There's a tarmac playground out the front with a house at the end as belongs to a chap called Brigadier Shaw. We'm always kicking our balls over the fence into Brigadier Shaw's garden, then trying to get them back without alerting his terrier. Up through the gate on the other side is the sports field, steep as the roof of a house.

Mrs Evans is droning on and I can't help picturing the day Father took me with him to top the fields and we used the horses. Some of the sons want tractors but fathers aren't planning to get rid of their horses, and those fields are too steep for tractors anyway. I was allowed to ride the horses, Shires, so tall and broad it's like straddling a tabletop. I could hardly walk at the end of the day, my legs were so bowed.

I don't notice Mrs Evans shouting my name. I hate Mrs Evans. She's got a face like a prune and so much food stuck between her teeth she needn't have eaten for a fortnight. Mother makes me clean my teeth twice a day and hits me with the flippy stick if I don't. Mrs Evans doesn't clean hers from one year to the next, and Mother bows and scrapes to her like she's Queen Elizabeth.

I use my fingers to lift my eyelids up, which is something I've worked out how to do to keep my eyes open at night when

I'm scared to fall asleep because of nightmares about the red stag in the wallpaper.

Mrs Evans doesn't like it.

God give me strength, she says. If you're not careful, I'm going to call your mother.

Well I don't want Mother to come marching up with the flippy stick, so I try to listen, but before long I start thinking about how it's nearly lunchtime up at the farm, and Father and Dave and Johnny will be taking their sandwiches into Granny's kitchen and Derek will be coming in and swivelling around in his secretary seat and kicking the pig under the table. I must've snorted or something at the memory of that old sow nearly carting the table off, because next thing I know Mrs Evans is coming at me with the ruler.

I fall sideways out of my chair and lie on the floor of the classroom, my mouth hanging open and my chair all tipped up, as if I'm dead. It's something I've been practising at home, and it works. Mrs Evans starts running.

Jean, she shouts, go and get Susan. Tell her her brother's fainted.

The rest of the children get up out of their chairs and cluster round and I can hear them giggling, and pretty soon I can feel my own laughter rising like a river in my chest and I can't hold it in.

I have to wait in the cloakroom for Mother to come and get me. The lunch bell goes and I'm hungry, but I'm not allowed any lunch. The other children stream out into the playground and run around laughing and kicking Jimmy Carson's new football and all I can do is sit there in the cloakroom and watch. I watch Jimmy Carson kick the ball into Brigadier Shaw's garden, and I watch them all gather at the fence and argue over who's going to go in and get it, only nobody wants to be the one who's got to squeeze under the fence and risk running into the terrier, or the Brigadier. He doesn't like children much. But the house is all closed up and there's no sign of the dog. I can see Jimmy Carson's brand new ball just sitting there on Brigadier Shaw's lawn, only a few feet from the fence.

I run out of the cloakroom and across the playground. I crawl under the fence, get the ball, and crawl back with it into the playground. I run towards the cloakroom with the ball clutched in both hands. I've got to get back to the cloakroom with the ball before Mrs Evans sees me, only before I get there Jimmy Carson sticks his leg out.

Snap. I land heavy on my knee.
   Crack. My thigh goes off like a bloody shotgun.
   They can hear me screaming on the other side of the village.

Mother's brought the flippy stick, only she doesn't use it. Ambulance man is leaning over me.

We gotta straighten your leg, he says. It's going to hurt.

Can't hurt more than what tis, says Mother, and she squeezes my hand hard.

Ambulance man pulls my leg straight and ties it to the other one. Hurts so much I faint, and I expect Mrs Evans thinks that's fair punishment.

They put me on a stretcher and take me back to the same hospital where I was hatched, only this time I'm on the men's ward. Royal Conservator's ward, East Reach. It's due to me being a brave little boy.

Mother comes to see me every single day, catches the bus from Winsford. Brings me a monkey that plays the cymbals. Yabba dabba doo booboo. Chink, chink, chink, chink. Still got him.

They put my leg in a pulley, only every time I turn in the night my leg pulls the cable off the pulley, and the weight drops, and down comes my leg, bang, and they have to start all over again.

1961

I go from plaster to bandage to calliper to crutches and after that I get a special pair of boots with a hole drilled through.

Jimmy Carson taunts me in school.
　Johnny says Derek says your father can't do farm work no more.
　He can do it, I says. He just don't want to.

My heart is breaking is the honest truth, and I haven't got words for it. Everything is breaking. Father is known all over Exmoor for his hard work, and for his hedges, which are second to none. Trouble is, not everyone wants to pay. Mother and Father are living from hand to mouth, and they don't have nothing in savings.

1963

I stand outside Edbrook cottages and touch the icicles hanging off the roof. The snow comes halfway up my bedroom window. The trees in the lane are coated in hoar. There are rooks and crows frozen to the branches.

It's a real heavy winter.

It freezes and freezes, then it rains and rains, then it freezes again. Then the wind gets up and blows the snow off the fields and onto the roads. It gets caught in the hedges, which are just dirt banks, with beech and ash and hazel planted on top. The drifts are higher than the cattle lorries. Father spends days shovelling snow to get to the farms. It's weeks before they get Gilbert Stanbury out of Withycombe.

The snow lasts until the new moon, and that's when Father goes to the hospital. They tell him he's got an ulcer and book him in for an operation. Mother says he'll be back up and running by spring, but Easter comes and he still isn't right.

He sells his tools, even his chainsaw, and gets a job working the petrol pumps in Minehead. People tip him if he wipes their windows when he fills up their cars, but even with the tips it's less than he got on the farms. And him a craftsman. He can turn his hand to anything, a myriad of things. He can lay hedges, build stone walls, harvest corn, make hay, do lambing, shearing, everything to do with farming.

Only now Father can't do farming no more, and that's how life starts to break apart.

1964

One, two, three, ding.

You got to pace it. Your footwork's got to be even.

I move across the field in the straightest line I can manage, heading for a pole at the other end. When I get to the pole I take it out of the ground, measure exactly three paces along, and put it back in the ground again. Then I turn around and go back, heading towards another pole at the other end of the field. I pace it as evenly as I can, then kick my toe in and scuff the ground. The poles mean I always have something to aim for, and scuffing the ground means I always know where I've been.

It's the week before we're due to leave Winsford for good. Father and I have been called up to help Derek one last time, and due to Father's condition, I'm allowed to pull the chain harrows on my own with the horse. First we have to plough the field, then we have to sow it with grass seed, then I get to chain-harrow it in. I'm eleven years old and it feels like the end of my childhood.

It takes us three days to sow the seed and in the evenings after we stop I can still feel the weight of the fiddle, slung over my right shoulder, and the box resting on my left hip. It's an expandable box, with hessian sides. You fill it up with seed and strap it onto yourself, only you never fill it right up, because it'll bust your shoulder. You got a little lever on the side that adjusts to how much seed you want per acre, maybe a pound an acre, and you got a little spinner in the front, like a cotton reel, only bigger. He'll spin one way, then back the other way, and you got a bow, and he's got a leather thong on him, come from the top to the bottom of this stick, and the stick's about three foot long, and he's fastened to the spinner on the front, and this bow comes down from the top, wraps around the cotton reel once, and all the way down to the end. You pull him in and the spinner'll spin this way, and you pull him out and he'll spin the other way, and that's how you scatter the seed.

The seed for the fiddles is stored in the link box on the trailer, which is hitched up to Derek's T20. I'm allowed to edge it down the field so we don't have to walk so far to fill up our sacks. I swing myself up into the open cab and move the gear lever to connect the starter motor, which fires up the engine. I let him roll forwards until Father or Derek shouts way, then I cut the engine, climb down out of the open cab, and refill my own sack with seed from the link box on the trailer.

Then on the fourth day it comes time to drive the horses over with the chain harrow, and it is I who walks behind them, chivvying them on to pull the harrows over the stony ground, pushing the seed into the soil. It's a warm day, and the field is a steep one, like all the fields around Winsford.

My leg is aching. Dr Winkworth did a good job, and when I'm wearing boots I wear them even, but from time to time my leg gets to aching, and it reminds me how things aren't right, and will never be right again.

Derek's son is pushing to replace the horses with another tractor. He reckons it'll be cheaper in the long run. But Derek says no, for as long as he's alive he'll keep the horses.

I'm glad. It's the joyest thing, walking behind the horses on a bright spring day, listening to skylarks overhead, and buzzards, and cuckoos, and sparrers clattering in the hedges. I can see Father in the distance, the sun behind him, watching me, and I'm thinking if I was a girl I'd weep.

When we get back to Edbrook cottages there's a line of dead chickens hanging off the washing line. Mother stunned them first, so they didn't swallow the blood, then hung them up by their legs, crossed their wings, banged them on the head and slit their throats.

Father's got a job working for John de Savery, down at Exmoor Woodcraft, and that's the reason we're leaving Winsford. It's too far to go every day.

They all end up there in the end. Mother gets a job there, then my sister gets a job there, then my Aunty Pearl gets a job there, then my Aunty Doris and my Uncle Derek and my Uncle Arthur. Whole bloody lot of them. Make gifts, that sort of thing. Trinkets.

1965

Brushford's a small village, and me and the vicar's son are the only two boys in it. We both attend the secondary modern, which is built on land as used to belong to Father's father. The farmhouse is still there, in the centre of the new houses, only it's empty now. Grandfather's long in the ground.

The new cottage doesn't have much of a garden, just a little patch of ground in front that never gets any sun. There's no room for Billie Jean so he got sent to the slaughterhouse. The only thing I can think of to do is go out shooting things with the vicar's son. His name is Horatio Willoughby Bourbon Busset, so you can see how we might not always see eye to eye, although we get on well enough when we're boys.

We get together at weekends and if there's nothing else to shoot, we shoot the streetlights, smash them to pieces. Everyone knows it's me and Horatio, but they never can prove it. We shoot Skipper Patterson in the arse from up in the church tower a hundred yards away. He's bent over gardening and he comes up to the gate that leads out onto the road and looks up and down to see where we are. Never thinks to look up at the church tower. I get air rifle pellets all over my hands and in my shins. Take them out with a penknife.

When we aren't shooting, we're catching baby rabbits and wild wood pigeons. I've got techniques Jim's taught me, and I pass them on to Horatio.

We wait until the young pigeons have hatched, then tie them down through the nest on a short bit of cat gut, which we get from the fishing shop. Not too tight around the leg or the leg'll come off. We tie them to the nest and let the mother feed them. She'll feed them and feed them and keep on feeding them, little porky pigeons, and she never can understand why they don't fledge. Then when we want one we go along and cut one out.

Rabbits we hook out of their holes with a good stiff bramble, size of a small thumb. We push them down the hole and twist them round and round and bring them up and have a look and see if there's any fluff on them, fluff and dried grass. If there is we know there's young in there, and we settle down to watch. You got to be patient. You got to watch for a few days. You got to have time on your hands. If they'm old enough they come out to play, and if they'm too young to come out we fish them out with the brambles. It's like fishing salmon out of the river. We bang them on the head, skin them, cut the legs and feet off and take them home to Mother, and she fries them whole in a frying pan. If you try to wring their necks like you would with a fully growed rabbit you'll pull the head off and it'll get mucky.

Rosanna Tarr rings me up. She's a very excitable, jolly person. GIRL. Year and eighteen months older than me, and taller. I'm a scrawny little thing.

Hare hounds meeting at Sandyway, she says, shall us pair up to run after them?

I borrow her hockey boots so I can run faster.

Saturdays we chase otters up the Quarme, watch the men pole them out from under the root plates of a tree, watch the hounds tear them apart.

1966

There's a robin comes and eats out of my hand. I go out every morning with a crust and just stand there in the front yard and he'll come and he'll take the bread right off my fingers.

There's usually a magpie watching from the hedge.
    Morning, Mr Magpie, I tell him. Some mornings there's two for joy, or even three for a girl, but usually he'm on his own. One for sorrow.

Father says tis my job to stop the jackdaws and magpies killing the songbirds and eating their eggs.

We go in spring, me and Johnny and Horatio, when the trees are bare, and we mark their nests, then we go around when they've got their eggs in and tip the nests out.

1967

The SS *Torrey Canyon* supertanker runs aground on rocks off the coast of Land's End, a hundred miles south-west of Exmoor, spilling an estimated 94 to 164 million litres of crude oil into the sea.

More than 12,300 individual bird casualties are recorded, including guillemots, razorbills, puffins, shags, great northern divers, red-throated divers, gannets, black-necked grebe, great skua and gulls.

2014

There's a rabbit sitting on the bank near the bird feeders, behind the cottage.

Go and give him a carrot, I says to Catrina. Go on.

I adore things like that now.

I watch her go out and up the bank and I try to square the part of me that loves animals with the part of me that loves hunting, but I can't. I never could.

She comes back inside.

Close the door, I says.

He'd gone, she says.

He'll come back.

She sits down at the table with her notebook spread out.

People generally pigeonhole you when you say you hunted animals, and the people who like to hunt, well some of them are bloody ignorant.

She'm scribbling.

I hunted everything from an egg to a fucken stag, I says, only I had to stop chasing after stags once they brought in mobile phones. They'd ring each other up and explain where he was and where he was going. Twasn't a fair fight any more.

Catrina stops scribbling.

I've never seen a hare, she says.

Well, I says, and I'm stuck for words.

Hares are precious now, but they were plentiful then.

Everything's precious when it's gone.

1968

I leave school when I turn fifteen and form a hedging business with Johnny and Horatio and a chap called Maurice Van der Ket. It's through him that I end up meeting Alfie Rudd. He's been in the Desert Rats, taken prisoner of war three times. All of us boys idolize him, and me especially. He's rumoured to be the best poacher on Exmoor. Father ketched him up at Knaplock once, stealing logs, so he'm none too keen.

We'm all riding in the front seat of Johnny's Ford Consul, and we got all the tools and the sheepdogs in the back, and our sign written on the side of the car, Hedging Contractors, only it's raining hard, and we've written him in emulsion paint, and when it rains the writing falls off and trails down the side of the car.

It's Friday, and lunchtime already, so we duck into the Sportsman's Inn in Sandyway to get out of the rain. Bill Floyd owns it, and he's a character like Derek. Their lands abutted each other until Bill Floyd sold up and bought the pub. Well, long story short, Alfie Rudd's in there, and we never get home until Monday.

After that Alfie and another old boy take to liking me, and we go weekends poaching salmon out of the Torridge or the Taw.

They don't have a car, only a motorbike and a sidecar. I go in the sidecar and the two of them are on the motorbike. They get the salmon out and kill them and chuck them in the field, then I come along with a bit of rope and tie the salmon to the rope. Then I put my hands in my pockets and walk along slowly, dragging the salmon behind me. If anyone happens to look over, all they'll see is a scrawny boy with his hands in his pockets, whistling.

The old boys stuff all the fish we catch in the sidecar and I travel home on top of it, come out the other end covered in scales. We poach the salmon out of the river that belongs to the hotel, then sell them back to the hotel.

Then one morning Alfie calls up my parents and tells them he's got me a position as an apprentice for Radley and Chandlers. Father says he doesn't want me going to work for no poacher, but Mother thinks it's a bright idea and Mother always wins.

1969

BUZZ ALDRIN: Beautiful view.
NEIL ARMSTRONG: Isn't that something?
 Magnificent sight out there.
BUZZ ALDRIN: Magnificent desolation.

1970

I put down my shovel, cross my arms, and pull my shirt over my head. I use the shirt to wipe the sweat off my face.

It's boring work, cleaning out the water tanks at the Tarr Steps Hotel, but better than the four days I just spent up at Old Man Chidgey's, digging a pit in a bog, because Chidgey's water's gone sour and he needs a new well. I worked evenings, too, cleaning up Chidgey's yard, which is all galvanized sheds with the roofs hanging off and cow shit seeping into the stream. It's awful work, although I do enjoy driving the old Davey Brown 990, with the little three-point linkage and McConnel digger attached.

Alfie stops digging and starts smoking, and so do I, even though smoking makes me sick.

First morning I go up there to dig the spring, I says, I seed some old caravans in the field by the stream. Old, old caravans they were. Only one of the caravans had legs. Well I get out and when I slam the door on the digger this old caravan starts rocking and then it starts walking across the bloody field. Then I seed the legs have got cow's feet, and bugger me a cow has gone into the caravan and gone through the floor and now he can't get out. I go up to Old Man Chidgey and I says there's a cow wearing one of your bloody caravans. Well they have to beat the caravan to pieces to get him out.

I'm trying to learn how to blow my cigarette smoke out in perfect rings like Alfie. I'm thinking it might help with the girls, especially Reenie, who works behind the bar in the Hare and Hound. I haven't had much luck at this point with girls. I'm taller than Alfie by half a foot, but I'm still scrawny. Arms thin as sticks.

Let us finish early, says Alfie, seeing as it's Friday. I knows a little spot. This time of year there's generally a salmon there.

We put our shovels away and go around to the front of the hotel and sit outside on one of the tables and drink two pints of cider each. Tisn't very clever. I stumble and sway on the path down to the river and my head aches.

Ah, says Alfie, when we reach the water. Bugger's here. Cut me a stick.

I cut him a piece of hazel and split the end. Alfie has a gaff in his wallet and a wire in his pocket and some extra wire and a string. He's always well prepared. He threads up the stick.

You'll have to watch both banks, he says. There's footpaths both sides.

Well he creeps in and gets down on his stomach and two

minutes later I see a couple walking towards us on our side of the river. I whistle, and Alfie gets up to his feet and chucks the stick down and comes to stand beside me, hands in his pockets.

We chat to the couple for a minute or two, how it's hot isn't it, and what a lovely evening. Then they wander back up to the hotel, and Alfie gets back down on his hands and knees and crawls back in. He isn't in there long before the same thing happens again. After it happens a third time he says ah fuck it, Tommy. T'int going to happen. We'm going to get kitched if we'm not careful.

He hides the stick under some brambles.

I'll come back in the morning, he says. Early like.

We go back up the path to the hotel and drink another three pints of cider. I say goodbye to Alfie in the car park. He gets on the motorbike and rides off and I'm supposed to follow, only I can hardly stand up, let alone stay on top of the moped all the way back to Brushford. Plus it's dark, and I know better than to get home and wake up Mother. I wander back down the path towards the river and curl up near where Alfie was trying to hook out the salmon.

It's not far off midsummer and I don't sleep long before the light gets in my eyes and wakes me up. I'm roaring hungry and I think well why don't I try and hook the salmon out. I'll take him home and have him for breakfast. I'll chuck the stick away and Alfie will never know it was me who got him. I think I'm being smart.

The salmon comes out easily and I wrap him up in my shirt and put him in my rucksack and carry him home like that on the moped.

Mother's sitting at the kitchen table with a cigarette balanced on the edge of the ashtray and a steaming cup of tea in front of her.
  Where you been?
  I went out and got you a salmon, Mother.
  I pull the salmon out of my rucksack and whip the shirt off him. He's a good-looking fish, even better in the smallness of the kitchen. Mother takes him and puts him in the sink. I have some breakfast, and after that I go up to my bedroom and I get into my bed and fall asleep.

We have the salmon for tea, and Mother is almost smiling when she tells Father how I've went out early in the morning and caught him straight up and brought him home. Father says he doesn't like me poaching alongside Alfie Rudd, but he eats the salmon same as the rest of us.

The Hare and Hound is busy and Reenie's behind the bar. Bernadette is the landlady and Reenie is Bernadette's daughter. She's got curly brown hair, and she wears red lipstick sent from America. Everyone and his dog wants to be seen alongside

Reenie, and I'm making my own way over to her when I see Alfie pushing his way towards me with a look upon his face. I know it straight off.

Well, he says, when he gets up to me. I looked in and the bugger was gone. Some bugger had ee, you know, did un, and I think tis you.

How you bloody work that out then? I says. Maybe he swimmed out.

Used the same stick did un. And I knows he wouldn't have swimmed out because the water in't high enough.

Alfie pokes me in the chest, and it causes me to stumble backwards.

It's all quiet like before a fight.

Some bugger as he, says Alfie, and there's only one that would know where that stick was.

He raises his massive fist suddenly only I see him coming and I duck, then I swing for it, and I catch him on the side of his face and he goes over. I knock him to the ground.

Back at the cottage Mother's waiting for me, hiding behind the door. Cottage is in darkness so I'm not prepared. As soon as I step inside she jumps on me, clings to my shirt, pulls my hair, scratches my face, scratches my eyes. Bernadette rung her up.

I don't know if it's the knowledge I knocked Alfie to the floor, the feeling of it, but I pick up Mother and waltz her to the settee. I shove her down and lean in close, my face right up against hers. She tries to shield herself with her hands, but I grab her wrists and pull her hands away. I'm shouting.

Don't you hit me no more.

Then the lights come on and Father's in the room with his chest puffed out and I'm scared, but I rise up like a stag and I turn to Father and I tell him, don't you come near me no more.

And he doesn't.

He tells me to get out of his house and not to come back.

2014

I need some technical assistance.

    I spin the chair round so I'm facing the room.

    Catrina goes and gets the old plastic water bottle Hope keeps under the sink, then she kneels on the floor and empties the contents of my leg bag into the bottle. First few times she did it she nearly bloody pulled the tube out of my leg. She'm getting the hang of it now.

She takes the bottle outside to empty it and I close my eyes. I must've fallen asleep because when I open my eyes she's just lying there looking at me.

    How tall are you? she says.

    Four foot and a tenner.

    How tall are you when you stand up?

    I don't stand up.

    When you used to stand up.

    Five foot nine. Short arse.

    Not that short.

    Standard.

1970

I get on the bus at Taunton, with my bag of masonry tools, my suitcase full of clothes, and three hundred quid stuffed in my shoe. I'm not streetwise at all. We hant got no streets in Dulverton. But I know enough to keep my money in my shoe, case I need to come home.

God almighty what a jump it is. Scratty-arsed little boy from Winsford, Dulverton, where everyone knows everyone's stuff, like what they had for breakfast, and how many carrots they grew, and then to get on the bus to London. I didn't realize there were that many people alive. I get a headache and it lasts for a fortnight.

I get off the bus and follow the instructions to Flanagan's pub, only it's full of Paddies, so I sit down on the pavement to wait.

I'm not sitting long on the pavement before someone else turns up, a big dark-haired boy, looks even younger than me, and he's also got a suitcase and a tool bag. I pluck up the courage to talk to him.

Are you going to ketch a bus here to Dover?

Yeah, he says.

Who's your agent?

Oh, he says, Peter Scheller.

Oh, I says, that's my agent.

Christ, he says, that's unlikely.

Let's us two work together then, I says.

His name is Jack Delaney and he's from Darlington, in the north, and he's even more scared than I am. I got to be a man. He's looking up to me.

After a few hours of waiting a bus turns up and we all clamber on. The Paddies are pissed as arseholes and they carry on drinking all night on the ferry, but me and Jack keep our wits about us. I'm not going to spend all my money on booze.

We travel together from Zeebrugge to Frankfurt and meet Peter Scheller in a big café outside the train station. He signs us up and gives us our tickets and tells us we got to go right across Germany. Hanover Minden's the place, he says. You got to report for work on Monday morning.

We travel for what feels like a fortnight, from one side of Germany to the other, and when the train pulls up at Hanover Minden we get off. The train pulls away and we look around and there's nothing at all. We're in the middle of nowhere. We walk down the road with our bags and our suitcases until we find a phone box, only I can't speak German and neither can Jack. Taxi, taxi, taxi, I says, only I can't understand what he's saying back. Hanover Minden Bahnhof, I tell him, repeating it over and over again, and lo and behold, half an hour later a taxi comes round to pick us up.

We walk in through the main gates, only before we know it there's a German leaning out of the window of his little hut, shouting at us. He's really shouting and shouting. Well I look at Jack, and we go over to the hut, and the German comes out and starts poking Jack in the chest.

Well I give him a push and I says, don't you ever fucking do that ever again, you bastard.

Next thing I hear is a little voice saying, now now, temper temper. I turn around and there's a dark-skinned guy standing there sniggering.

Can you speak English? I says to him.

Yeah yeah, he says, I can speak English.

Well, I says, tell this ignorant fucker this is where we been told to report.

Now now, says the man, temper temper.

His name is Otto and he's put in charge of us.

First he says we got to break up concrete using jackhammers. They got it all marked out.

No, I says to Jack, after five minutes of it. Bollocks to this. They can shove it up their arse. I come over here to lay bricks not bloody hack up concrete. That's not my trade.

I march back to the planning office where Otto has his desk. Get your bloody labourers to do it, I says.

Otto laughs. Okay, he says. Come.

The next thing he wants us to do is carry bricks and mix mortar for a bloke who's repairing a wall within the confines of the factory. Jack starts running around, but I just stand there with my arms folded.

You want me to *labour* to him? I don't spect he can even *lay* a fucken brick.

Otto laughs again. Gets right under my skin.

I got Father in my head, chest puffed out. Got him in there all day long and all night too. I don't know what to do with him.

No, I says. No way.

No?

No.

Otto drives us to a site outside the factory, some sort of school, where a team of men are lined up on a piece of scaffolding, building some huge great long single-block wall. Must have been ten of them, all German, with Turks running around below mixing mortar and loading bricks and operating the cranes. We climb up the scaffolding and take our places on the line, which is held up in the middle by a tingle, to keep it level.

I know I got to do quarter bond, instead of half bond like they do in England. I reach down and pick up the mortar, spread it on, reach down again, pick up a block, slap en on, only before I can slap en on the German next to me has slapped on one of his own.

I clean my block off and try again, thinking I'll be quicker this time, flip flop, only by the time I turn around he's done it again.

Jack, I says, are you having the same problem as me? Every time you put your bed of mortar down, the bloody German next to ee just chucks a block down willy-nilly?

Yeah, he says.

Well, I says. Fuck em then. Let them get on and build this bloody wall.

I climb down from the scaffolding, and Jack follows me, and we stand there like that, with our arms folded. Then Otto comes along and screams at us, and he's going with his arms like a windmill.

No, I says. Don't shout at me. Otto, Otto, Otto.

It's no good. So I bend down and pick up my level and hold it up against the side of his face and I says, if you shout at me one more fucken word I'm going to knock your fucken head off.

So, ah, says Otto. Vas is dis fucking? Vas is fucking? Huh? Sfucking? Why he, vas dis fucking?

Next thing Otto gets one of the bosses to come down from head office, and we'm thinking they'm going to send us home, but instead they take us to the planning office and show us some plans. They got some real special brickwork, Otto says, a long line of face brick, got to be completely smooth, and there are portholes in it.

It's a terrible difficult job. Every joint has to be measured. Every joint has to be the same.

What do you reckon? I says to Jack. Can you do it?

I'll be all right, I says, I can do it.

Yeah, says Jack, I'll have a go. And him only done a six-month crash course in bricklaying.

Otto is smirking and sniggering. It's our last chance and he thinks we'm done for. But we finish the job, and quickly, and after he's been to inspect our work he invites us into his little office. He gets the bottle of schnapps out.

Come, come, my friends, he says, and he offers us overtime with the concrete gang, pouring concrete, and that's us set up.

We work all day, have a meal in the pub and then a drink, then we work all night. We go to bed at four o'clock in the morning and we're ready to start again at seven.

1973

I'm leaning in right underneath the hopper, with the steelwork leaning on top of me, and it's a big construction, and matey who has the rope attached to the hopper is letting him down, and he's still letting him down, and he's coming down and he's squashing me, and they're all shouting. Oy oy oy! Stop stop stop! And when he stops, he stops quick, and then he goes down even further. And all these bloody Germans are trying to hold this thing up and it nearly bloody crushes me.

Ding!

That night I dream I'm walking upwards, through a gate, through a patch of woodland, out into an area of bracken, and there are flowers on the high ground and I wake up sick with longing.

Jack, I says.
What? he says.
Nothing, I says.

Then the frost gets so deep everything grinds to a halt. Otto tells us to go on home for Christmas and come back again in March.

I board the train with Jack and we say goodbye in London, same place we said hello, and I'm thinking I'll see him again soon enough, only I don't. I never go back.

I meet a bloody woman instead, and get married.

7

Global cement production has increased more than thirtyfold since 1950, and almost fourfold since 1990.[7]

1974

At first, Father and Mother are happy enough to have their son back, but soon Father is on at me to be up and working like he was at my age, even though I've got thirteen grand in the bank, so I rent a caravan on the outskirts of Brompton Regis, and set about spending the money I saved on booze and horses.

We gather in the evenings at the Poltimore Arms. Me and Reenie and Bobby and Johnny and Horatio and Rosanna Tarr, and I tell up my tales like a bloody peacock, like a young stag just growed his first pair of horns. Most of them have never been to Exeter, let alone Germany. I catch Reenie looking over, and I'm aware that I'm not a scrawny little thing any more, and she's got engaged to Jimmy Carson, and he's none too pleased.

Well, you know what they say about pride comes before a fall. I had plenty of pride and I done plenty of falling.

I'm up near the caravan one afternoon in February when damn me I see a bloody hind. It's gunshot range and the wind is in my favour, so I go in. I gut and gralloch her there by the side of the road, and skin her out, and sling the carcass over my shoulder and carry on. Only then I hear a car coming.

I can remember when Brompton Regis had a police station and a Saturday bus to the market and a forge. Even in the 1950s there was such a thing as a morning rush hour at the forge. Farmers needed horses shod and ready to work in the fields before the sun rose too high and it was too hot for the horses to work. But tractors put an end to blacksmiths. The forge at Brompton Regis got turned into a house in 1979. Bus stopped in 1968, with the advent of cars.

My caravan is on the outskirts of the village, and I don't get many visitors, so I'm not prepared, but I know the lie of the land, and I jump in over the bank to hide, only as I hit the bottom of the ditch my gun goes off.

Damn, I'm thinking. That's given my game away.

I wait until I can't hear anything, then I drag myself and the bloody hind back up over the bank. It's dusk and I can't see much, but I can see the car is still there. It isn't a farmer's car, so I keep on walking towards it, with the hind slung over my shoulder. Don't have much choice at that point.

She's strange-looking, the girl who gets out of the car, dressed like a man with hair so dark it looks almost black, and cut off sharp at the chin. Wearing blue jeans cut tight, boots with heels.

You must be Tommy, she says, in a strange, flat accent.

Well I am, I says. But who are you?

Gracie, she says.

Clive's cousin. I had heard she was over from somewhere in America, Wyoming, or Montana, somewhere like that. Clive works at the petrol station in Dulverton, and he never stops talking about his aunt and cousins in America, how you can buy land out there for next to nothing and go chasing cattle on horseback. Clive's known for drinking and fighting. I can't hardly believe this girl is Clive's cousin.

I've come up here to find you, she says, and she reaches up and tucks the front of her hair behind her ears. Her hands are small and her nails are all bitten down. Skin pale like stretched fabric.

I get a chill all down my arms. Goosebumps, like I seen a ghost.

Why've you been looking for me?

Well, she says, bold as a fox, Clive says you got horses, and I was hoping to ride one of em.

I catch her eye then, and her eyes are green, not brown, and she stares right back, and it's like the whole world stops. Like a thick snow's come down and everything's gone silent.

She's wild as a hawk.

A week later we're travelling along in my Austin Maxi and we find an owl in the middle of the road, stunned. He's only a baby. Little fluffy snowman. We get out and he's hopping along the ground, trying to get away, and swivelling his head round to look back at us, and that funny little chirp they do. I used to be able to imitate it, but I hant got the breath any more.

She picks him up and hands him to me and he puts one of his talons right through my little finger. Excruciating pain, but I don't howl. Drive all the way back to the caravan like that, with an owl stuck to my finger. We get him home, and the next day I'm thinking, I don't want to raise him in an enclosed environment, I want him to be able to be free. So I build a scrap of a nest – they never have much of a nest – in a saw shed down at the bottom of the field, and I place him in there.

In the morning I go down and see a chap in the village who used to rear pheasants and hatch them and he gives me a supply of day-old chicks. I feed the owl on day-old chicks.

I go out every night and drop him a day-old chick, and in the daytime he'll sometimes come inside the caravan and sit on the shelf.

Week later a man comes round to see whether it's worth us having a television. I'm sitting on one side of the table, and the owl is perched on the shelf behind me. Matey's on the other side of the table, and we'm talking about this and that and the other, and he keeps on looking up, and the owl is perfectly still, very serene. Then all of a sudden he turns his head round, and matey nearly jumps out of his skin.

Bugger me, he says. I thought he was a stuffed one.

We got to take him back to his nest at night because if we don't he'll fly about everywhere, crash into all the furniture and the chinaware.

Gracie's living in the caravan now, along with me, and she wants a dog, so I drive to Wolverhampton and come back with a pup, all white and fluffy with blue eyes. My first ever Samoyed pup. Costs me ninety quid. I call him Sam, and the pair of us join forces to adore our mistress. We only got eyes for Gracie. She'm the only person in the world.

Mother and Father are smitten too, especially Father, who's in the habit of thinking everything I do is wrong. There's a constant barricade of how I'll never make anything of myself, because I got money in the bank and I'm not out all day like he used to be when he was my age, working on the farms.

But he loves Gracie. Everyone loves Gracie.

Except Reenie, but I put this down to Gracie being Reenie's polar opposite, and to Reenie being jealous, because she had my eyes fixed on her for so long.

Gracie's father is Native American, which is where she gets her cheekbones and her liking for horses. We're already hatching plans to move out there, get a farm of our own in the mountains, get some land.

Then the day comes for me to go back to Germany and I don't go. I get a pang when I think about Jack but I can't tear myself away.

I lie on the bed in the caravan with Gracie on my shoulder and the owl on his shelf and Sam in his basket, drinking whisky and smoking and dreaming and telling up my tales, and it serves me right what happens next because I let everyone down.

1975

I cut the cake with a chainsaw. Mother and Father are present, and my sister Susan, and all my aunts and uncles, and all my cousins. We have the party in the Hare and Hound in Brushford, and the wedding in St Mary Magdalene's Church, Winsford.

The bells are ringing when I stumble up and I can hardly believe any of it is true. Gracie, the wedding, the fact of my life.

I always loved being in the church with the bells ringing. It was one of my favourite things. There were a few of us kids that learnt to use the bells. There was a bell-ringers' club and we used to practise one night a week. I wasn't very good because I wasn't heavy enough. But I used to go any rate, every Sunday morning, put spit and polish on my hair and toll the bell. There was a little bell just for me. I used to toll the bell and get the people coming to church.

And now I'm standing at the front of the church in my hunting suit, watching Gracie walk up the aisle towards me.

It's Clive gives her away, all dressed up in a brown jacket and even a tie.

   He's a big bugger.

   You better treat her well, he says.

   I'll bloody kill anyone who tries to harm her, I says, and he puts his hand on my shoulder and squeezes it slightly and him six feet six inches tall.

We spend the night in the Lion Hotel in Dulverton.

I remember it all, the smell of her, the feel of her fingers on my jawbone, on my shoulders, on my neck, on my chest.

   How would you like to die? she says, when we're finished.

   Not much, I says.

   I don't think a plane crash would be so bad, she says.

   I bury my face in her hair. It smells of lavender.

   I hear drowning's nice.

1976

It's the hottest summer since records began. We're renting twenty acres and I got my three horses on it and then I buy a shepherding pony for Gracie and we rent another twenty acres for making hay. Then Gracie wants sheep, so we get sheep. I suppose I spoil her, but she spoils me too. Buys me loads of gifts.

Then she says she's going back to Wyoming to find us a farm and I want to go along with her but she says I ought to stay behind and mind the stock.

I drive her to Heathrow and it's the middle of the night and we stand by the entrance and kiss on and on and I'm running my fingers through her hair, holding clumps of it in my hands.
Let me go now, she says, or I'll miss my bloody flight.

Getting married and buying the horses and renting the land has used up all the money I saved in Germany so I'm working my bollocks off building a slurry system for a gentleman farmer.

I based the design on something I saw in Holland when we were passing through on the train. I drew it on the back of a fag packet and showed it to the farmer.

First I have to excavate it, six foot deep and three foot deep underneath. Then I have to build it and it takes three hundred blocks. It earns me a reputation all over Exmoor, and keeps me occupied right up until it's time to drive back to Heathrow and pick up Gracie.

I get there early and wait in the pickup with Sam. I've got the advert with me and I'm reading it over and over even though I already know it off by heart. I could recite it in church if I had to.

Twenty mountain acres north of Sun Valley.
 Total price 9,900 dollars. 1,900 down. Ten years at 115 dollars per month.
 208 683-2124. 208 263-6885. Mobile 1020.

Turn it over.

Dear Thomas,
Picked this up in a cowboy pub near Ketchum. Cheap, isn't it? Drove the Buick from Ma's place through the Rockies to Idaho where we changed the clocks to Pacific time so we're a whole day behind you.
 So remote here you wouldn't believe it. I keep on saying, if only

I had the jeep. The motel we're in is a kind of mountain lodge, you'd love it.

Two dollars fifty to stay in a cabin. Saw a brown bear yesterday. Hardly seen a person since we crossed the border.

Her flight comes in only she isn't on it.

I wait eight hours for the next one, like a bloody idiot, then I turn around and drive home.

1976

Next morning a boy comes with a telegram.

In Alaska stop. Please sell stock stop. Not coming home.

I call Clive from the payphone in Brompton Regis. He says Gracie was due to meet a cousin of theirs near Vancouver, but she never turned up. She's met a man, he says, a fugitive, and gone with him to Alaska.

Farmer comes by with Sam in the back of his Land Rover.
 Killed three sheep, farmer says.
 It's like someone flicked a switch.
 Sam knows something's up, he's looking at me, pleading. I go inside and get my shotgun and shoot him in the head.
 I didn't mean that, says the farmer, and he takes his cap off and twists it in his hands.
 I didn't mean it neither.
 I wait for him to drive off then I fire my gun straight up into the air to keep people away. Telephone won't stop

ringing so I rip it out of the socket and leave it on the wall by the caravan.

I bury Sam in his favourite spot by the wall at the back of the field. Bawl my bloody eyes out. Reenie comes round when she can't get me on the telephone.

Tell her to bugger off and leave me alone.

Owl's formed a habit of sitting in the ash tree outside the caravan and squeaking at me. Every bloody night. Come on, Mother, give me some more food. I've taken to going to bed with a torch. Every time he squeaks and kicks up a fuss I shine the torch out through the window to make him quieten down.

That night I wait until the owl starts squeaking then I go out with my chainsaw and cut the tree down.

One of the long-haired saboteurs opens a gate to confuse the hounds.

You close that fucken gate, I says.

No, he says.

You close that fucking gate or I'm going to drag you through that fence. I'm going to turn you into chips.

I'm wild.

I get a fox for a pet, take the place of Sam. Dig him out. Sleeps all day and restless all night. Keep him indoors in a box and take him around with me in the pickup. Put him on a lead, take him for walks.

2014

Catrina is scribbling in her notebook.
    Was he on his own when you dug him out?
    No, he was a cub.
    What happened to the others?
    Shot em.

2009

Hope comes to the hospital.
 I got bad news.
 She picks my hand up and holds it in hers.
 Spike.

That night is a long, dark tunnel.
 I dream I'm on a skateboard with no wheels, just a square bit of timber sliding down the tarmac and out towards the edge of a huge quarry. I think I'll be able to turn around but I can't. I end up right out over the edge, screaming.

I wake up in floods of tears, then the nurses come to lift and turn me and then I got to eat and it all starts again.

It's only dogs who'll die of a broken heart. Rest of us keep on living, or try to.

1977

I sell the stock and turn to forestry.

I start with cordwood, lengths of timber already felled and stacked on the ground, which I cut and chop and sell as logs. I borrow the money for a new chainsaw and a circular saw, and soon I've saved enough to buy an old tractor.

I work until the sweat is running in a river all the way from the top of my head down to the tips of my toes.

I buy a second-hand Land Rover and a second-hand trailer, and eventually I buy a winch and a cradle for the tractor and then I'm all set up.

I'm a fast worker and I build a reputation for myself by word of mouth. Soon I'm offered standing timber. Nothing is too hard for me. I lug my tools in on foot if I have to, cut like crazy all day long. I stack the cut wood into piles and cover them with a tarpaulin. It's a matter of pride to me to deliver the logs dry.

Then I'm offered pulpwood, and it's a step harder again, and a step more money. To keep the licence from the pulp company, I got to guarantee a given amount each month, be it softwood or hardwood.

Alfie Rudd drops dead from a heart attack, and his son Andrew's only sixteen years old, so I take him on as an apprentice. It's the least I can do.

In the mornings we load the cut wood onto the lorry, 14 tonnes of it, all by hand. There are very few lorry cranes around in that part of the country. In the afternoons we go back and fell more wood, cut it, and drag it to the access point for the following morning. Sometimes, when the driver has a clear run to Chepstow, he'll offload and get back in time for five o'clock.

I got space for another one, he'll say, jumping down out of the cab, and I'll load it up again, by myself this time, because Andrew has gone home. I'll load 28 tonnes of timber, by hand, in a single day.

1978

It's a shock to learn Gracie's back in Dulverton, holding wild parties, trying to buy all my friends.

First time I see her in the street it's like someone kicked me in the stomach.

Hey Gracie! I shout, only she pretends not to hear. I see her colour up. She crosses the street and goes into a shop.

I suppose that's when I know she's really gone.

We'm still married at that point, but it isn't long before her mother turns up at the caravan with divorce papers and a bunch of solicitors. Killer. I've failed at something again. It smashes my confidence to pieces.

They go round saying, this is what we're having, and they take a load of things aren't theirs to take. I don't care. Then Gracie comes to the stable where I got my horse and tries to take him too.

He's suffering from mud fever and you have to treat it a certain way. You got to lightly whisk the dry mud off with a brush. I'm keeping on top of it with the vet and a few other witch doctors. There's a routine, so Gracie knows exactly where I'll be.

She turns up at the stable while I'm in the middle of whisking off the mud and before long we get right into it. I throw the brush over the stable door and it hits her in the eye. I don't mean to hit her at all. It's an accident. But Gracie always did bruise up easily, and her eye turns black, and before long Clive's on the warpath and half of Dulverton are crossing the street to avoid me.

Clive has a reputation for being a vicious fighter, and people are shitting themselves, thinking there's going to be a bloodbath.

I fight Clive and I send him home worse for wear and then I can't stop fighting. There are people who think it's exciting hanging around a guy who's strange and fightable. They hang around and egg me on. I walk into a pub or a barn dance and if someone appears to be staring at me I go right up and clout them.

I fight everyone who looks at me funny, and everyone who dares me, right up until I get my face smashed in.

1979

These various habitats, some sheltered, others wild and exposed, attract the many animals and plants which make Exmoor such a wild and fascinating place both to the naturalist and the lover of the countryside. Here the wild red deer have survived for a thousand years and more, and there are other species of deer as you will discover in this book. The Exmoor pony, though partly domesticated today, is another old native along with the fox, badger, otter, and dozens of other creatures. Over 100 species of birds nest every year, and many others migrate through or spend the winter months here. Most of the thousand flowering plants are also part of ancient Exmoor and link us today with past, unrecorded ages.

N. V. Allen[8]

1979

It's Exford show and I'm smoking wacky baccy round the back of a caravan with half a dozen Australian sheepshearers.

Then someone starts in.

Lift up the caravan and tip en over.

He can't do it.

I can.

I put my shoulder to the bloody caravan, and I start to pick him up, only before I can tip him over I hear a screeching and a hollering coming from inside. I drop the caravan and a girl comes screaming out. Turns out it's the ladies' toilet.

Not only that but one of the sheepshearers is the girl's boyfriend. They all set upon me for trying to tip over the trailer with a girl still inside of it.

It isn't very pretty.

I wake up eight hours later in Taunton hospital and I can't see a thing.

I'm blind, I tell the nurses.

I've got blurred vision for three days, and when that stops my heart starts aching. It's a stabbing pain on the left side of my chest, shooting pains down the left side of my chest.

Can't have been easy, what I put him through.

Twasn't easy what he put me through neither.

2014

You look uncomfortable, says Catrina, when she comes along to give me my rice cakes and Marmite. I wish I could fix that.

Me too, I says. If I thought you could I'd ask you to do it. Never had my legs ache so much in my bloody life as since I lost them.

We go down the hill and along the lane beside the river.

There's no wind. The only sound is the sound of the chair and the sound of the water.

See there. I bring the chair to a halt. Get one of them leaves off the ground.

She bends down and picks up a leaf.

Pass it here.

I'm trying to train her up a bit when it comes to trees. It's amazing how much knowledge she's lacking, seeing as she's never lived in a city. Although I suppose there aren't too many trees in her part of Cornwall. Generally speaking, though, people don't know anything any more. We've lost most of our knowledge.

Look, stems are flat, feel them. That's why they make that rattling, rustling noise.

What is it?

Aspen.

Jesus's cross was made of aspen.

I stop so she can open the gate and close it behind us.

My favourite tree is oak, she says. What's yours?

Downy birch. They call her the Lady of the Woods, even though catkins can be male or female. Short-lived, eighty years and she'm ready to fall over. Second favourite is cedar of Lebanon, then Scots pine.

We pass the ash on the corner.

See that, I says. They'm saying we're going to lose ninety per cent. Twill be like the elms.

1980

Every farm on Exmoor has towering elms for shelter and to get wood for furniture and coffins. Every hedgerow has an elm or two or three. There's plenty of work bringing them all down. Sixty million have to be felled countrywide, and there must be more than a million on Exmoor.

I buy myself a harness and quickly earn a reputation for being a tree surgeon. I'm self-taught and I've only done the basic stuff, but I know trees. Elm is hardest to cut, blunts every blade.

I'm hired by J. H. Baker to work with the team who are felling up at the Royal Ordnance Factory, near Bridgwater. It's a highly secure site where they test explosives. We got enough trees up there to last us for months, and it's through John Baker I meet Heather.

She's got two boys the same age as his, and she sometimes comes out to the pub with us on a Friday night, when the boys are with their father.

She lives in one of the houses on Grandfather's old farm, and she attended Dulverton school like I did, only I didn't know her then. She's a couple of years younger than I am, a happy person, cheerful soul, always laughing and cooking for people. Polar opposite of Gracie.

It's like walking into a ready-made family. I take to going round in the evenings and sitting at the table with her and the boys. It's like I'm their dad.

We keep it at friends, and I try not to let my mind wander into deeper territory.

I'm wary.

1980

The female earth was central to the organic cosmology that was undermined by the Scientific Revolution and the rise of a market-oriented culture in early modern Europe. The ecology movement has reawakened interest in the values and concepts associated historically with the premodern organic world. The ecological model and its associated ethics make possible a fresh and critical interpretation of the rise of modern science in the crucial period when our cosmos ceased to be viewed as an organism and became instead a machine.

Carolyn Merchant[9]

1981

After I leave the compound I'm offered paid work felling all sorts of species and sizes of tree. Sometimes I process them for milling too, and it's through this I manage to save the money to buy myself a Ferrari. He's an alpine tractor, articulated, with equal-sized wheels. I put a little winch on him, and he's the ideal tractor for first and second thinnings.

All goes well until I agree to do a job for a firewood merchant I know in Winsford.

I want to winch some timber down, he says to me in the pub one night. I just got to get it out.

He explains to me where the timber is – at the top of one of those steep fields by Halse farm.

All right, I says.

Nobody else will dare go up there in a tractor with a winch attached, but I don't care. The steeper the better. I can name my price for these kinds of jobs, and I like going up to Winsford.

The following weekend I turn in through the gateway and drive the little tractor up to the top of the field, keeping my foot on the accelerator.

I wait for the pain in my heart to gather and pass away, then I swing the tractor round and reverse him in as close as I can to three massive branches that have come off one of the old beeches in the hedgerow. I notice that someone has already taken the tops off the branches and that gives me pause. Without the tops on, the branches could roll and tip the tractor over. I select the smallest of the branches and hitch it to the tractor using the winch. I drive as slowly as I can back down the field and we make it without a mishap. I unload the branch and set about cutting it up for firewood.

I drive up the field for the second branch, and here my luck starts to run out. As I'm turning at the top, the lower-side wheel drops down into a rabbit hole, and the tractor tips over. I manage to jump off in time and I stand and watch him roll off down the field. He's going so slowly I'm tempted to run and step in there and hold him back, but I decide tis probably not a good idea, and after five rolls he does come to a halt.

I walk into the village and find Johnny. He comes and helps me get the tractor back upright again, then I drive him home and spend the rest of the day checking him over. I go over everything, looking for damage. I even change the oil. The tractor seems fine, so the next morning I set off once again over the hill to finish the job.

I'm about halfway down the field with the final branch when it all goes multi-pear-shaped. It's a real heavy branch, and the tractor is rearing up with the effort of pulling, in spite of the steepness of the field, which I find strange. I'm thinking I must go steady, and steadying up, only then there's a loud bang, and the next thing the tractor has folded in half. There must have been a crack in the articulation, between the front and back wheels. The tractor is still driving, but he won't steer. The hydraulics and the steering are linked to the back wheels, with the engine and gearbox over the front, only he's swivelled in the middle, and the seat has flipped right up under me. The timber is vertical, twenty feet of it, rearing up behind, and I'm still plunging down the field at top speed, with no steering and no brakes. I cling on, afraid to let go. I close my eyes when he approaches the flat part of the field and starts racing towards the thick brush hedge separating the field from the lane behind.

He crashes into the hedge, ploughs right through him, crashes over the lane and crashes into the river before he stops.

1982

Let's the two of us go and climb Crib Goch, says Horatio, after I finish telling him the story about the tractor.

I haven't seen him for years. He's been living in Exeter, got a wife and a job as a solicitor.

Well, I says. I've been along Crib Goch in summertime, and I got no desire to do it again, especially not in the wintertime.

But he manages to talk me round.

We go up there two days before Christmas and it's a blinding bright day. Takes us four hours to drive up, and when we get to the mountains there's a lot of snow – too much – and there's been a lot of wind, so there's a lot of cornices, which is when the wind blows the snow up the side of the ridge and then the snow keeps on going beyond the solid material, out into nothingness. It's terrible dangerous. You can't see the edges of the mountain, and if you go too far out they'll break off.

I don't think we should go up, I tell Horatio. Not like this. Not with this snow and this wind.

Well I'm not going home without climbing the ridge, he says. If you won't come I'm going on my own.

Bloody stupid.

We rope ourselves together at the bottom of the first climb. I should have been in front and him behind. Silly bugger is heavier than me. But he had to be out front, only then we get up onto the ridge and it's blowing a bloody gale up there and he falls through the cornice. I can see him up in front dangling and kicking in mid-air.

There's a saying in the Alps about those types of ridges. If your mate falls into Italy you gotta jump into France.

Well the only thing I can think of to do to stop the pair of us crashing to our deaths is dangle myself off the other side.

I have to rely on my ice axe to keep me steady while he tries to haul himself back up again. It's the scariest thing. It's bloody awful. He's climbing up, using me as the anchor, and he's bloody bouncing, and I'm shouting stop fucking bouncing, and the wind is so strong he can't hear a thing, and we should never have been up there in the first place. One slip and game over.

We make it up to the top and I'm bloody shaking and by the time we're walking down it's dark.

Remember that daft bloody bank manager who came up here with a bottle of whisky and no clothes?

Horatio is all skipping and jumping after his fall.

No, I says.

Yip, says Horatio. Killed hisself.

I'm not listening to him. It's a full moon and all the peaks are lit up. Beauty. Absolutely beautiful.

My old man's got wind of a cottage, Horatio says later that evening, when we're home safe and settled into our table at the Hare and Hound. You interested?

Not really.

What'll you do when you want to bring your wife home? Won't look too good bringing her to an old caravan.

I ain't got no wife, I says. Not any more.

I don't want to mention Heather. I'm waiting to see how it beds in before telling people. Don't want them all talking.

Well, says Horatio, we might as well go and look at it.

1983

Cottage is lovely, surrounded by woodland, oak and hazel and willow. Used to be part of an estate, but they tore the estate house down and replaced it years ago. There's another cottage across the road but they've got nobody living in it. It's a holiday house.

I sign the lease and a month later I load up the truck with all my things, ten years' worth of treasure, and drive it over to the cottage from the caravan. I lay it all out on the kitchen floor. My slots and gin traps, my rabbit wires and weaponry, my Damascus-barrelled guns, my hammer guns, my crossbow.

I got the crossbow because I thought it would be nice to have a killing machine for deer. He's got a 210-pound draw, and I been angling to see how far the bolt will go, only I never wanted to try him out when I was living on the outskirts of a village.

I take the crossbow out onto the doorstep of my new home and I fire a bolt straight up into the air.

Eventually it occurs to me he'll have to come down again somewhere, and me standing there like a bloody idiot looking up at the sky. I go back inside and look at the clock and wait ten minutes, then I go back outside again, only there's no sign of him anywhere. I haven't got a clue where that bolt come back down. I never heard of anybody wandering around Exmoor with a crossbow bolt in their head, so maybe he never did.

1984

I love living up there by myself, far away from the nearest village. I can get my bugle out and prance and dance up through the woodlands whenever I want.

When it snows I go down and light a fire in the woods and cook sausages and sleep out there in my sleeping bag. It's like going back in time a hundred years.

I drive to Swindon to get another Samoyed pup, call him Sizzle, and when he's big enough I build him a sled and get him a harness and have him tow half a bag of coal up the driveway. Then I find a pair of skis in the second-hand shop in Dulverton. I've never been on skis before in my life, but I get Sizzle into his harness, and get myself into the skis, and stumble to my feet. I manage to hold on long enough to have him pull me across the driveway and up into the neighbour's field. Luckily it's only a gentle slope because Sizzle won't go in a straight line, no matter how much I cuss him. There's nothing I can do about it. I'm cussing him and cussing him to get him to go straight, but it don't make no difference. He's darting all over the place, sniffing out fox and badger and what have you. It's all I can do to hold on.

But I do hold on and we make it up to the top of the field, and that's when I hear the sound of an engine approaching. I turn around and see Heather's car coming slowly up the driveway towards the house.

Get on home, I say to Sizzle. Go on home.

Well now he'm only too happy to go in a straight line. He charges towards the cottage, dragging me behind him, and as we're approaching the driveway I can see Heather's car edging close, sliding a little bit on the ice, and I'm thinking shit, Sizzle's going to pull me out right in front of her car.

Well he does exactly that, and there's nothing I can do to stop him. Heather slithers to a halt and watches from the driver's seat while Sizzle drags me across the driveway and into the yard. I let go of the reins, crash into the hedge, kick out of the skis and emerge out the back of the hedge, sauntering along with a massive grin on my face.

Well, says Heather. I have never in my life witnessed anything like that. You wait until I tell the boys. I'm taking that to the grave.

That's when it starts, and it's not like it was with Gracie.

The feelings are there but they're different. More like friends. Like deep friends. More like brother and sister. I can tell Heather anything.

Only then she says, why don't I move in with her.

Twould save on money, she says, and I'm round at hers all the time anyway.

I like it the way it is, I tell her, with me going down to hers weekends, and her coming to me Tuesdays, and the rest of the time having the cottage to myself. I got all my things and where would I keep them at hers? I got a Bultaco Pursang mark 5. Scrambling bikes. Kayaks. Canoes.

1985

Scientists with the British Antarctic Survey announce the discovery of a huge hole in the ozone layer over Antarctica.[10]

6

There has been a fiftyfold increase in the production of man-made substances since 1950.[11]

2014

Never in a million years did I think I'd be lying here on a Sunday morning having some woman come in and wash my cock.

Get the Black and Decker out, I says to Sian, one of my carers, give me a trim.

I tried it myself once with the hair trimmer, but I didn't have the strength to hold him up, nicked myself on the balls.

Hope was rattled.

What were you thinking?

Are you telling me off?

Yes.

It's a medical day. Cor blimey they wipe me out.

They trained up my bowels in the hospital. Trained them up so I go every other day in the morning, between half past eight and half past ten. Got to have suppositories the night before, and in the morning, and more often than not by the time the carers get in I'll have already had a good start. It's called dysreflexia. If you don't poo the toxins will kill you, and even more so when you're paralysed from the diaphragm down.

Carers have to clear up the bed pad then manually evacuate whatever's left. I've got them trained up. Talked them through it. They have to tell me what they feel and see and I got to explain what to do. Sweep or hook. I lie on my side and I got a certain amount of sensation like that, inside. It isn't that different to a machine. You got to understand the workings and you got to pay close attention.

Takes me one and a half hours to get up, and two and a half on a medical day. They got to get me washed and dressed, get the sling in place, hoist me into the chair. That's the worst bit. I've been lying flat for hours, and then my organs suddenly shift. Blood pressure drops and I get light-headed, like going up on a fairground ride, like when you tip yourself back and up suddenly. Head rush. Takes two of them to get me into the chair. One of them to operate the hoist and the other one to guide me into place.

1986

Tractor's a rare beauty. Regular four-cylinder Fordson Major, but with a frame that's been stretched to accommodate a six-cylinder Perkins engine. I pay for him with cash I saved from various jobs, and he's pricey. I reckon he's worth it, only it emerges during my first week of ownership that he's a bit of a bugger to start when he's warm. It isn't too much of a problem. I just have to keep him ticking over while I eat my sandwiches.

Well this goes on until one day I've had enough of the noise of the engine. I want a bit of peace and quiet. Bugger it, he'll be okay. I turn him off.

I eat the sandwiches Heather made for me and then I lie down on my back in the dirt with my cap over my face. I'm trying to solve my dilemma. Heather is agitating for an answer and her patience is wored out. I got to choose between her and my cottage, she says.

I can see her point. I've started to imagine how maybe she could be my next wife, if she wasn't married already, which she is, only he isn't nowhere to be seen. I can picture my life going down that path. It's a happy picture. Happy enough.

Only I know if I give up the cottage I'll never get him back. I took him on with a long lease, but the manor has changed hands since, and I happen to know my new landlord is itching to get me out and put the rent up.

It's Friday, and I've got a lot of work to finish before the weekend. I haul myself to my feet and brush myself down and go over to try to start the tractor. Course he won't start. He squeaks and he whistles, but nothing happens. Battery is flat. I wait a few minutes. Sometimes if I wait for the squeak of the solenoid to die down he'll start all of a sudden, but not this time.

I decide to go and get the battery out of the other tractor, my old Ferrari, which I've kept for easy jobs that don't require much pulling strength. He's parked out on the other side of the wood. I gather up my spanners and trudge for half an hour through the heat, and when I get to the other tractor I think hang on, it's bloody daft to take the battery out and lug it all the way back through the wood. Why don't I just drive the old tractor up?

So I drive the old tractor back through the wood to the new tractor, and while I'm driving it I think hold on, it's bloody daft to go to the trouble of taking the battery out of the old tractor. Why don't I just drive the old tractor up behind the new tractor and give the new tractor a push? I can jump out of the old tractor and jump into the new one and save myself a whole lot of time and effort.

I leave the old tractor idling and climb up into the new one. I press the ignition and put him in top gear, thinking he'll turn over easier like that. I jump back in the old tractor and nudge the new one forwards, just a little nudge. Should be enough to get him going, I'm thinking, only I left the throttle wide open.

I nudge the new tractor with the old tractor, and bugger me, the new tractor goes off like a bloody greyhound down the track. I jump out of the old tractor and run as fast as I can, thinking shit, I'm never going to catch the bloody thing. Well I do catch him, only then I realize I can't get up over the back of the new tractor, because the anchors are on the back. I'll have to try and get around the side of it, jump up that way. Only I can't.

I'm hopping and skipping along the side of the new tractor, wondering what the hell to do, until about ten yards from the gateway out of the woods and into a farmer's field, I manage to reach in and jam my hand on the clutch and press it in. I use my other hand to grab the fore-end loader and I let it pull me along, skidding along, until the tractor stops.

Well now I've got a new problem. I've got to get him out of gear. I lean in and feel for the gear lever, only I can't reach him, not with my right hand, and my left hand is on the clutch.

I expect you know what's coming, and it's true. I only bloody swap hands, don't I. Let the clutch out like a bloody idiot. Well tractor knocks me over and runs right over me, and it's not like it's mud either, it's a hardcore track. I'm lying face down, and he runs over my legs and my hips and my stomach.

I come round a few minutes later. I remember what happened. Bugger me, I'm thinking. I'm dead. Only I can hear the tractor buzzing. Christ, I'm thinking. Bloody tractor must've come along with me.

I manage to sit up and look out over the field and there he is, doing a lap of honour. He must have clipped the bank when he crashed through the gateway, and the wheels must have turned a fraction.

I haul myself to my feet and I'm a little unsteady. I watch the tractor lumber towards me up the field. It's now or never. I run and I jump up onto the side of the tractor, and with my right hand I press the clutch, and with my left hand I pull the gear lever, and the bastard bloody thing rolls to a halt.

My head is thumping and the birds sound like jackhammers, and I'm thinking, this is going to hurt. I've just been run over by three tonnes of tractor. My calf muscle is twisted round, almost like it's been tore off the bone, and my stomach is heaving. I slide to the ground and wait for the pain, but he doesn't come as badly as I'm expecting, so after about half an hour I get myself up and back to the Mini pickup, and I drive myself to Dulverton.

The woman in Netherton Stores comes out from behind the till. She's seen me struggling with my box of frozen goods, and it's not like Thomas Hedley to struggle with anything, apart from women.

What on earth you been and done?
Run myself over with my tractor.
You going to be all right?
Yeah bugger, I says. I'll be all right.

I gather up my box and I get myself back to the pickup with it and I get myself home. I think about going to Heather's instead, but she'll most probably still be at work, and I don't want the boys to see me in that state. I get myself home and run a deep, hot bath, and I soak in there for three hours, until the water is stone cold, and then I try to get out, only I can't.

I stay like that, in the cold water, until Heather turns up to find out why I never answered the phone.

Oh my God. Jesus Christ. You got to go to hospital.

Nah bugger no, I says. I'll be all right. Just get me out of this bath.

You're going to hospital, she says, and I don't care what you say.

I've got clots all over my legs. Doctors give me warfarin to thin my blood, make me wear special tights. Tell me I won't be working for a while.

Yeah bugger, I tell them.

They give me a pair of crutches and I sling them over my shoulder. Nurse comes running up.

Mr Hedley, she says. Use the crutches.

Next morning I phone up the DSS and the lady on the other end of the phone listens to the story and then she says, well I do find that hard to believe, but I spose I gotta believe you, and she gives me an appointment to go down and sign on, only I don't end up going. Miss my appointment with the physiotherapist, too.

I wait a week then call up Andrew Rudd. I figure he owes me a favour or two.

Andrew, I says, you're never going to believe this. I've run myself over with my own tractor.

Well, he says, I do believe it coming from you.

I can hardly bloody walk, I says, but my top half's working. I think I can drive the swing shovel.

All right, says Andrew. Come up if you want. We'll get you in there somehow.

I haul myself into the Mini pickup and drive it up to the site and I do the whole thing in second gear, because I don't have the strength in my leg to press the clutch.

The labourers pick me up and sit me in the seat of the swing shovel, then they pick up my crutches and stuff them in behind the seat. They put my lunch on the door handle and Sizzle between my legs.

We'll come and get you out at five o'clock, they tell me, and they set me going. Bloody nuts.

1986

An accident at Chernobyl nuclear power station, in northern Ukraine, causes the largest uncontrolled radioactive release into the environment ever recorded for any civilian operation.[12]

2014

Hope's chin's resting on my head. I'm operating the computer with the stub of a pencil I've got strapped to my wrist with Velcro. We're both looking at the picture on the screen.

He's a Mercedes Sprinter, long wheel-base, and he'd make an excellent camper, only instead of a steering wheel he's got a joystick. I use the pencil to go through the pictures.

There's a ramp at the rear end operated with a remote control, so you can get yourself in and out without needing assistance. There's clamps on the floor for the chair. It's got automatic transmission and the stick for steering and there's buttons for lights and going backwards and indicating and all of that.

Looks like a bloody aeroplane cockpit, I says.

You've got all that experience driving cranes and winches and tractors. Should be a doddle.

I don't know.

I think you should get him.

He's terrible expensive.

You can afford it.

I picture myself driving to market and doing the shopping and driving to the livestock auctions. Going down to the pharmacy on my own instead of relying on Hope to do everything. Picking her up and bringing her back if she wants to go out with her friends.

Well, I says. I suppose I can.

I'm up and running half past ten the day they deliver him. I put my Pink Floyd CD on the hi-fi in my study and waltz around the room. *What do you make of it all?* I put on Cher. *You'll be crying over me.* Turn him all the way up and spin around the study in my chair, until Hope shouts at me to turn him down.

The delivery driver works for the company that did the adaptation and he offers to give me a tutorial. He shows me how to operate the ramp, and he has me start him up with a button I got hanging around my neck, and drive him backwards and forwards a few metres. I'm wondering if he's thinking I'm too bad in my condition, worse than what he's used to, but he doesn't bat an eyelid.

I wait until he's all the way up and down the lane and out of sight before I have a go at operating the ramp, getting myself into position, getting the chair to clip into the bolts on the floor. This is the first challenge. It's hard to get him lined up.

I go out every morning after that, to practise, and by the end of the first week I can take him out of the parking spot and up the drive and back down again, a distance of around ten metres. I edge him up and I edge him back down again and after an hour of that I'm done, exhausted.

It's a full month before I leave the confines of the driveway and set out down the lane towards the village.

1988

After I run myself over Heather stops talking about me moving in, and another year passes, and now it's hard to bring it up. Something between us has changed, only I can't put my finger on what exactly. I assume it's related to me choosing my cottage over her.

Christmas comes and goes and I don't go down. Her husband's back in the area, and he doesn't want me there with the boys. He's the jealous sort.

I don't go down New Year's either. Don't go anywhere. I sit in my living room, on the sofa I got out of the Red Lion, the one Winston Churchill sat on, and I look at all my treasured possessions arranged on my walls and surfaces and I'm thinking to myself how none of it matters. I've made my decision. I want to go along with Heather, and if it means giving up my cottage then I'll do it.

I find a piece of paper and an envelope, and I write a letter to my landlord, giving notice on my lease. I post it the following morning, which is a bank holiday. I think if I wait I'll change my mind.

The following evening I put on my new boots and set off for Dulverton. I'm going to tell her my decision. I'm excited. She's been angling for it so long.

Stars are raining down.

I get a hell of a shock to hear her say she's changed her mind. Husband doesn't want another man living with her boys, she says. In fact, she's been thinking we ought to cool it, on account of him being back in town.

Takes me a few days to get over the shock.

Oh well, I think, when I'm able to think, that's my dilemma over. I've still got my cottage.

Only then I remember the letter, and I think shit, I'm going to be fucking homeless.

I sit there and try to think my way out of my predicament, but in the end I only have one idea. Landlord hasn't come up yet, which means he probably hasn't received the letter, with it being Christmas.

God almighty, I say out loud. I'm going to have to hold up the postman.

I go down before work, when I know they'll be having breakfast. I creep through the woods and hide outside my landlord's house. I know the post goes on top of his washing machine, because I've seen it when I've been round there to take him the rent. I can hear them all in there, the sound of plates and cups. I tiptoe in, steal the mail, take it home, sift through, nothing. I go down again next morning, steal the mail, take it home, sift through, still nothing. Third day I go down, steal the mail, take it home, sift through, find my own letter, take it out, burn the rest, and my landlord never knows.

I'm on my own for a long time after Heather. It's not that I blame her. I've got no confidence left. I've been burnt twice, and I don't feel much like putting my hand back in the fire.

But we settle into a friendship, because of the boys, and it's lucky we keep that.

If it wasn't for Heather I'd be dead.

2015

Hope's got a scarecrow in the competition, along with all her friends. They've really gone to town on it, put them on display in the hedgerows. Bloody scarecrows adorning the fields for miles around. Hope made hers green, in the shape of a spider. Scare the Devil he would, never mind the crows. Takes third place.

We say farewell after the judging. I watch her walk away, then I lose her in the crowd and that's that. It's the first time we've been apart longer than a few hours since I come out of hospital, and she's due to be gone ten days.

Catrina's talking all sorts of rubbish and I know she's only trying to help but I do wish she'd shut up.

People are clustered around the roped-off areas, watching the animals go round. I'd like to watch the heavy horses, only I don't fancy my chances. Oftentimes people will come and stand in front of the chair and they'll think I'm sitting down and they won't notice I can't get up.

It's the same with pavements. We'll go into town and people'll be standing in groups chatting, and they'll look at me and then they'll look away and stay where they are, and I've got to say excuse me.

Sun is beating down and I hant got my hat.

Shall us go on home? I says.

We push through the crowds towards the place where I left the van. I'm wored out by the time we get back to it, but with Hope gone in her own car, it's down to me to drive the pair of us home.

You wrote a book yet?

No.

Well I wrote two or three, you ought to write one.

That pisses her off. T'int easy, she says.

Well, I says, nothing worth doing is.

I pull off the roundabout onto the A470. I used to love driving, but it scares the shit out of me now. I don't like the thought of Hope driving on the motorway all the way to Devon, and then London.

I got to slow down then and get in behind a line of cars and vans lined up behind a tractor.

Bloody shit.

It's like the M5, says Catrina.

ONCE UPON A RAVEN'S NEST

We get home and I park my chair by the windows.

Catrina goes outside and comes back in.

What are you doing?

I'm tapping my head and then tapping my jaw, I says.

If there's a reverb or an echo, then I know there's infection. It's like the sound of an engine. You learn to listen and you can hear there's something wrong.

1988

The safe boundary of 350 parts per million of CO2 in the Earth's atmosphere is passed.

1991

It's coming up to Christmas and business is flourishing. I've got Andrew Rudd working for me, and another chap, called Lindon Flodometch, and I decide to take the boys to the pub on Christmas Eve. We have a raucous night. I'm paying for everything, and God bugger do we drink some stuff. Get absolutely pounded. Drunk as handcarts. Leave my Austin Maxi in Dulverton and take a taxi home to the cottage. I can hardly bloody stand up. I take some of the pills I got for the pains in my chest – doctor thinks it's a hiatus hernia – then I scramble into bed with all my clothes on and pass out.

Christmas Day I feel rough.
    Heather's husband's off the scene again, and I'm due for lunch at hers – just friends. I drag myself out of bed to ring her up.
    Cor, I says. I got no energy.
    Well, you had a bloody skinful, says Heather.
    I won't come down to lunch, I says.
    Come teatime, she says. The boys'll want to see you.
    All right. I'll come teatime.

I get back into bed for a couple of hours, and when I wake up I feel ever so slightly better. Good. I'll get up and have a snack. That'll usually sort out a hangover. I eat a couple of crackers with Stilton, and I swallow a couple of Gaviscon, thinking they might settle me down, only instead of that they make me bring it all back up again. I collapse on the floor with a dreadful cramping pain in my chest. I'm gasping for air, crawling towards the door, thinking to get outside. Sizzle is following along and I can see he's worried, but there's nothing I can do. We're both a bit disconcerted. I don't know what the bloody hell is going on. Warning bells are going off, only before I can respond I pass out.

When I come round I'm lying in a pool of my own vomit and there's something heavy sitting on my chest. It isn't Sizzle. He's lying down next to me on the floor. I drag myself to the phone and ring up Heather again.

I can't come down. I'm bloody worse. I passed out. I've been sick every bloody where. I'm weak. I got this pain in my chest. I don't know what to do with myself.

I'm coming up, says Heather, and she does come up, and she takes one look at me and calls the doctor.

The usual doctor's off for Christmas, so it's the locum who comes up to the cottage. I'm propped up in the armchair, my legs straight out in front of me.

You've been drinking over the Christmas period, he says. Take some more of your Gaviscon. You'll be all right.

I take more of the Gaviscon, and this time he stays down, and the pain in my chest eases a little bit.

You'll be all right now, says the locum, and he gathers up his bag and leaves.

Maybe it's true, I says to Heather. It's the worst hangover I ever had, but I'm thirty-eight years old. Maybe it's my age.

Only Heather seen me on dozens of hangovers and she says no, she isn't leaving me alone.

Good thing she doesn't.

Midnight she calls the locum again, and he comes back up to the cottage.

Right, he says, tutting and grumbling. I'm going to examine you. Hmm, he says, listening to the stethoscope. Hmm, ahh, yes, hmmm. Well, he says, I'm going to send you into hospital. I don't want to be called out again in the early hours of the morning. But you'll have to make your own way in. They need the ambulance for emergencies.

My Austin Maxi is still in Dulverton, where I left her the night before, and anyway I can't drive, so Heather drives me to Taunton in her car, ancient bloody thing, with more rust on him than paint. I can see her white knuckles shining in the moonlight.

It's my fifth visit to Taunton hospital and I still manage to go in the wrong entrance. I'm out of it. Walk around the whole building with my legs collapsing.

Receptionist says she's spoken to the GP.

Locum, I says, my own GP knows me, and he would have known to tell you something was up.

The receptionist picks up the phone and calls someone in another part of the hospital. I can hear them arguing over where to put me and I'm falling over, legs buckling underneath me.

Chuck me in any bed, I says, I don't care. I need to lie down.

The space in front of my eyes is dark and filling up with little stars, like exploding fireworks.

Heather supports me to the ward, takes my full weight, and she'm only small. When I get there I collapse on the floor. The nurses scoop me up like a wet flannel and get me into bed. After about ten minutes a doctor comes in with an ECG machine. He plugs it in and wires it up and nothing happens.

Oh, says the doctor, this machine is broke, and he goes away again.

An hour or so passes and I start drifting away. I'm having trouble breathing and I can't see anything, only darkness. I know the doctor isn't coming back. Nobody's coming. They've given up on me. I'm dying.

I struggle into a sitting position.

Drag me out of this place. I ain't gonna die here.

I fall backwards, and when I come round I can hear voices. Heather ripping into somebody in the corridor.

Then a different doctor comes in, with one of the nurses from before.

Where's the ECG for this man?

Oh, says the nurse, it's broken.

Well, says the doctor, go and get another machine.

So they go and get another machine, and they plug it in, and wire it up, only before the doctor gets halfway through he stops.

I can see very well what your problem is, Mr Hedley. You've had a major heart attack.

1992

To overshoot means to go too far, to grow so large so quickly that limits are exceeded. When an overshoot occurs, it induces stresses that begin to slow or stop growth. The three causes of overshoot are always the same, at any scale from personal to planetary. First, there is growth, acceleration, rapid change. Second, there is some form of limit or barrier, beyond which the moving system may not safely go. Third, there is a delay or mistake in the perceptions and the responses that try to keep the system within its limits.

Donella Meadows [13, 14]

1992

Doctor says I got to think about how I'm going to spend my retirement. Signs me off heavy work, and me not even forty years old.

I've got fluid around my heart, he says, pericarditis, restricting it from beating, because I was left so long without treatment.

Must have seen it written on my face.

Mr Hedley, he says, do you know how lucky you are?

5

The ozone hole still exists today, but it has stabilized, thanks to the Montreal Protocol, which came into force in 1989.[15]

1992

I see the camera in the window of the Red Cross in Dulverton. It's an old-fashioned camera, even for back then. I buy him on impulse, thinking perhaps I can go around taking pictures of kids at gymkhanas and selling them.

It's Heather's idea for me to go and do an A level in photography at Dulverton College. I'm the oldest bastard in the class by about a hundred years but I don't worry. That's when I start learning the habits of deer, not to kill them, but to take their pictures.

It isn't easy. If you'm stalking deer, you've got to be dead still. Hinds'll usually give the game away. They'll sense you're there. Their sight's not that good, but their sense of hearing and smell is excellent. They can sense you even when they can't see you. I've had them lie down and wait. You can't move. You can't even blink. She'll graze a little bit, and then, if she turns, you know she'm going to move on. But generally if you'm staying still they won't notice you. Except when some other bugger comes crashing up through the woodland making a load of noise, and then I'll be cussing because I've watched them half an hour to an hour and got them all settled in, only for some other bugger to come and move them on. I never use soap, because soap is scented, and deer pick that up quicker than anything. Same with toothpaste. Not that you got to go out stinking like a pole cat. You got to *wash*. The scent is quite sweet, to deer.

It gets so as I'm able to predict exactly when and where the stags will drop their horns, and I know the age of the deer from the size and shape of their horns. I pride myself on working out which animal the horns belong to. I know their habits and habitats, what they eat, when and where they sleep. I know their movements on different nights, and how and why their movements might vary. It gets to the point I can predict to the day when the first pair of antlers will drop in the Minehead area, and when the first pair will drop in the Exe valley area, and it might be four days later, something like that.

2015

Catrina is lying stretched out on Winston Churchill's sofa.

Christ, I'd do anything to stretch out like that. Stretch my legs out. Move my legs.

She'm looking at the horns I got hung up on the wall.

Why do they all look different from each other?

Well, I says. Most people say the bay point comes at five year old, but if the deer is feeding off really good pasture he'll develop that point sooner. Sometimes he'll even grow that point one year and nothing on the other side. Next year he'll swap sides. Also if you feed deer mineral food he'll grow a colossal pair of horns, and each year they'll get bigger and bigger. There's so many theories behind it, and you can't look it up in a book, too many exceptions. You got to get out there and learn it.

I take hold of the walking stick I keep near the fire, for opening and closing the vent, and I sort of like wave it up at one of the pairs of horns I got mounted on the wall.

See, I says. He'm getting old. His horns are getting smaller in size. But the coronet at the bottom, that stays the same size. These stags are known as autumn stags and the smaller ones would be known as spring stags.

Trey point, bay point, brow point. See? Bay point he attains when he'm about five year old, point defines from male deer to a stag.

She'm not listening.

Do you have to kill the stags to get the antlers?

No, horns fall off because of testosterone. They grow them for mating season, then they shed them. Grow them every year until they're about nine or ten, then they start to wither and get smaller. You have to go up on the moor and look for them. It's good fun. No, it's more than that. Anyone who ever collected a pair of horns will tell you the same thing: if the bug bites, you don't stand a chance.

1993

I find my first pair on 15th March. I'll never forget the date. It's cold for the time of year and I've been out for hours. I'm on my way back to the car, which I parked up a forestry track off the main road, only I see a stag about to come into one of the fields on my left. I work out I can get into the field one of two ways, through the contractor's yard, or through the main farmyard. Well I don't want to set off all the dogs, so I cut across just behind contractor's cottage, and I make it over without being seen, and I get settled in behind this big old oak tree.

I'm not there very long before the stags come out, half a bloody dozen of them. I got to climb the tree to keep my scent above them. I sit up there and watch them for hours and I'm hoping they'll graze away from the tree, but it gets darker and darker, and they only go and lie down. It must've been about midnight before I finally come out, and my legs stiff as boards.

I can't go through either of the yards, because of the risk of setting off the dogs and triggering the lights. The only other option is a long walk around, up to the narrow part of the river, and then back on down again. I know there's a pole across the river on the other side of the field, put there by the keepers, so I set off, and I find the pole, and I slither across it, and I get almost to the bank before I slip and fall arse over head into the river.

Turns out that's my luck. I scramble up the bank the other side and that's where I find the horn, the first one, and I only find it because I'm down on my hands and knees, and then, lo and behold, the other one is just a little bit further along. My first pair of horns.

Only that isn't the end of the story.

I got a new car, second-hand but new to me, he's a Citroën XM, nice car, with self-heating seats, and I don't want to get the seats all wet and muddy, so I decide I'll strip my clothes off, all except my T-shirt, and sling them in the back. It's only about three mile before I'll be home and I can get into a nice hot bath. I drive off, only when I get back to the road there's a vehicle coming down the road towards me. I wait for him to go past so I can pull out, only when it goes past I see it's a bloody cop car, and I'm thinking, oh fuck it, here we go, middle of the night, and what am I doing pulling out of a forestry track?

Two of them get out of the car, and they come around to the side, and I roll my window down and I says sorry officer, before you start, I'm not going to get out of the car. I got no clothes on.

He shines his torch.

Oh, he says, where are your clothes?

So I tell him the story, and explain about the horns. Taking antlers is like foraging for berries, so it's permitted. Taking deer is poaching, and that's theft. I've taken so many deer in my life I'm surprised they can't see it written all over my forehead. It's the first time in my life I've been driving home at three in the morning without a gun and a bloody carcass in the back of the car.

Well, the policeman says, we had the car reported as being in a strange place, and we did think it might be stolen.

I haven't got my licence on me or nothing, but I says to them, you can follow me home, and I tell them where I live. They open up the back of the car and forage around and they find the horns and my clothes in a wet pile, and they open up the boot and my gun isn't there.

Best thing you can do is get yourself home and get changed.

Well I get myself home and washed and changed and next morning I'm up and out in the fields at four o'clock, with my camera on my shoulder, looking for horns.

Bloody nuts.

I go right up to the top of the moor and listen to the sound of life stirring in the valleys down below. The first pheasant crowing, and then the deer are on the move and I'm just standing there watching them. Pure happiness.

Then autumn comes and I get obsessed with listening to the stags roaring. It's a craze on Exmoor at the time, and I get right into it. I make up microphones with parabolic speakers to record them, and I buy a second-hand video camera, and I make a carrier so I can carry it all.

I listen for the stags and then I imitate them and they come galloping up. I never knew anyone get gored by a stag in the wild, only hounds out hunting, but there are times I struggle to hold my ground, they look so fierce. The challenge is to get the clearest recording you can.

People are always asking me what the hell I'm doing, charging up and down with my heart the way it is, but I know to rest for a while in the afternoons.

I'm trying to learn what I can about my heart, so I can heal myself and get rid of the tablets I'm taking. I know how to read the signs and I got ways to alleviate the pain.

1993

When I'm not following the stags around I'm out at the coast with my kayak, and then one day I walk into a shop in Croyde and buy myself a longboard, and that's another bug that if he bites you don't stand a chance.

Old boy called Mervyn Vasey takes pity on me.

I see him every day sitting on the beach soaking up the sunshine like a retired lobster. He used to have a hand plane, and he'd spend hours out in the water with it. He'd be out in the middle of winter, out in all conditions. Then arthritis got the better of him, so now he can't get in.

Well he sees me flailing about and one day he says come here, boy, I'll teach you to surf.

Takes us two hours to get to the water's edge, and him hobbling along seventy-eight years old. I'm embarrassed, because he makes me do rabbit jumps, and he's shouting no, you're twisted, no, you're dragging your feet.

I won the championship one year, English surf kayak championship, men's division, but surfing is different. You got to be balanced. You got to be in tune. Everything's got to come together.

The locals don't like it when I enter the line-up, at least not at first, but it isn't long before Andy Harrison and me take to liking each other.

Andy's a fireman from Saunton and wild in the water. When we become friends the old boys have to lay off a bit. They have to accept me.

For the first time in my life I'm content.

I've got the stags, and my camera, and surfing, and once a week I take myself to the cinema in Tiverton.

I go on my own and I sit at the back and I watch whatever they got showing.

It's like time has slowed and opened out and I can do all the things I never had time for beforehand, because I was always working.

2010

Hope says why don't we go to the coast for a picnic so we load me and the wheelbarrow into the van and drive over to Aberystwyth and out to the end of the harbour and there's half a dozen people surfing. I can't take my eyes off them. I watch the waves breaking on the sand, curling over one after the other after the other, all the same and all different, and I can feel myself inside them, being pulled along, lifted up, going along, carving and turning.

Then I'm back in my chair and my chest is closing over and I can't breathe and it's like being dragged under and I'm drowning.

4

Half of all coral reefs have been lost since the 1950s, along with 63 per cent of their associated biodiversity.[16]

1993

There's a pub as you approach Bude from the southern end. It's halfway down a steep hill, if my memory serves me, which it often doesn't these days. The road heads out, then it heads down towards the pub, quite steep, then it reaches a sharp right-hand bend.

They've put bales on the bend to stop us skating off the edge of the cliff, but other than that there isn't much in the way of health and safety.

Bude downhill run. I've never done it before, but Andy convinces me. Says it'll be the ideal way to celebrate my fortieth birthday, and I'm stupid enough to believe him.

I get myself a skateboard in the shop at the top end of town, a one-pound bargain, shaped like a teardrop, and I carry him up to the start of the race.

It's hellish fun watching. There are people just crashing out everywhere. They try to take the racing angle, and go wide and smash into the bales. I watch a few, then I set off on my one-pound bargain, and he's bloody quick. First try I make it round the first bend and fall off. Second try I get all the way down, and all the boys are cheering. Andy Harrison and Big Al and all those boys. It's a master challenge and I'm proud, only my heart is beating irregular, like he does when I overextend myself, so I sit down on one of the benches they got there.

Next thing I know there's a woman sitting next to me.
Well you burnt up the show, she says.
I did really.
It's one of those warm nights when you can sit out. You only get two or three in the course of a year.
She's got a ring on her third finger, so I can see she's married, but she doesn't mention it, and neither do I. Says her name is Gillian. We go on up to the pub, and the others let on it's my birthday, and she goes in and comes out with drinks for everyone, and we all sit there drinking until Andy says the lads are moving on, and I tell them to go along without me.

It's like an omen. The letter G. I should have run a thousand miles. Instead I get her to tell me her place of work, and, lo and behold, it's the dentist in Taunton. Well I never had so many teeth needed checking. Haunt the place. Eventually I ask her out, and that's when she tells me outright she's married, husband's a vet.

I know I should give it up, but I don't. I can't.

1994

Hen pheasant you can shoot anywhere and bring her down, but cock pheasant you got to shoot in the head, in the eye. I often sit and listen when they go to roost. They'm noisy buggers, cock pheasants, and if I listen I can hear where they've all gone. Hen pheasants are silent, just a little peep peep peep when they get up to flight.

I'm taking one or two pheasants a week, and maybe a couple of rabbits, just for my own use, and because Gillian likes rabbit stew. I'll slip into the woods with my torch and my handgun and my white dog, would you believe it, only as soon as he hears the shot he'll be on his way, wrong direction.

I've stopped going after deer. I love them too much now to shoot them. Only one night I'm up on Anstey Common and I happen to have my blunderbuss with me, ten-bore, and I see a hind, and she's been injured, so I bring her down.

I blast her out, gut her and gralloch her there in the woods, bleed her out, drag her back to the Mini pickup and load her in. There's blood still leaking out of the carcass, and it gets all over the back of the pickup.

I drive home and call Heather.

I got some nanny goat, I says.

That's what we call it. Never call it deer or venison, always nanny goat. Case anyone's listening. Heather's always on the lookout. She's always phoning me up.

You got any nanny goat?

No I bloody hant.

All right, well give us a shout if you get any.

She comes to help me skin her and cut her up, and we're packing the bits away ready for the freezer when there's a knock on the door. I go to the door and it's my bloody landlord.

Well he's been coming down hard on poachers lately and I don't want him to see what I got, in case he thinks I took it off his own land, so I keep him dancing on the doorstep for as long as I can, hoping he'll say his piece and bugger off, but he sort of like pushes himself inside, says he's got something important to tell me.

I make as much noise as I can coming up the hall towards the kitchen. I rattle the door loud, so Heather knows we're coming in. He goes first, and I hang back, waiting for the fireworks, but nothing happens, so I go on in after him. Well bugger me, Heather's sitting there on the sofa, nice as pie, and no sign of any deer. None at all. I'm thinking shit, where's it all gone?

Then landlord says his piece and I forget all about the deer.

It isn't until he's gone and I've got over my surprise enough to pour us both a glass of whisky that I remember to ask.

Where's all the bloody meat?

Look in the sofa, Heather says, in behind the cushions.

I get up and have a look, and bugger me there's all these cuts of meat just stuffed in behind the bloody cushions. She hadn't even had time to put it in bags. It's got a sticky membrane on it, venison. Soon as you chuck it down, it picks up all the hair and stones and stuff. There's more cushion stuck to that bloody carcass than there is left on the sofa, I says, but Heather isn't listening.

What're you going to do?

Wash it under the tap, I says.

No, she says, about what he said.

I don't know.

Well, she says, twould be a shame to have to let this place go.

Well, I says, you've changed your bloody tune.

I don't sleep much that night. I get up and go downstairs for a glass of water, and end up drinking the rest of the bottle of whisky on my own in the kitchen. I'm sitting in there and I'm looking at it all laid out, all my knick-knacks, my gin traps, my sofa from the Red Lion that Winston Churchill sat on, my first three pairs of horns. I've had them mounted by Patrick, who runs the gun shop in Dulverton, and hung them high up on the walls. I've got dozens of boxes and bowls I've carved out of wood over the years. Dozens of objects and all of them precious. It's my life, all laid out, and I been there so long I can't imagine it somewhere else.

I wake up the next morning with a bloody hangover. Phone's ringing. Gillian. We've got a rare day together and we've been planning it for ages.

We normally only get a few hours here and there. The recharging of the batteries. That's what we call it between ourselves. She'll come to the cottage and we'll walk up onto the moor or we'll sit inside by the fire. We're always talking. We leave the threads of our conversation hanging and pick them up again next time we get together.

Sometimes she'll meet me in Saunton or Woolacombe and she'll walk up and down the beach while I'm out surfing. Her husband thinks she's gone down to visit her sister in Cornwall.

Gillian knows exactly what she wants, and she knows what I'm offering, and she knows if she wants it or she doesn't. She listens to my suggestions, and she's interested or she isn't, and there's no excitement and drama. She knows her mind, and her mind is in the right place. It's all right looking at these young fit girls, but at the end of the day a woman's in her prime at forty.

We drift closer and closer, and there's no hurry in it. We're waiting for her kids to grow up so she can leave her husband and we can walk off into the sunset. Time is full of sweetness, the sweetness of making plans, and waiting for them to come to fruition, and knowing they will. I got so much faith in me and Gillian. Never question it for a second.

Gillian lights a cigarette and holds out the packet and I take one and light it, even though I haven't smoked for ten years. Gillian can blow smoke rings as good as Alfie Rudd when she's in the mood, only she isn't in the mood. She's smoking like a baby sucking milk.

Well, she says. Are you going to tell me what's going on?

Tis the cottage, I says, this cottage, and I feel even worse then, because it's the place we can be together. I know she loves it too.

Looks like I'm going to be moving on.

Where to?

Well, I says, I don't know. My landlord comes round last night and tells me he's putting him on the market.

What?

That's right.

Well, says Gillian, there's only one thing you can do then.

You got to apply for a mortgage and make him an offer. He's got to give you a good price.

 I been signed off, I says. Who's going to give me a mortgage?

 Well, she says, twould be worth a try.

I know the owner of a tea shop in Dulverton and I know he's short of staff, so I call him up first, and he offers me a job, only my very first day the girl that's meant to be doing tea-making and salads and what have you leaves me on my own. Before long these people come in and order a pot of tea. I take it over to them on a silver tray.

 Here you are, Madam, I says.

 A few minutes later she beckons me back.

 It's too strong, she says.

 So I make another pot and take it over.

 Here you are, Madam, I says.

 Few minutes later she beckons me back.

 It's too weak, she says.

 Well I've had enough.

 Make your own fucken tea, I says, and I stalk out.

Bank got to see proof of earnings so I swing back into forestry, in spite of my heart, and that's when everything changes again. The bank agree to a mortgage, and the landlord sells me the cottage and I'm back to the life I had before.

It kills everything when the cottage comes up for sale.

I hardly see Gillian. She can't get away evenings or weekends, and I'm back to working all hours of the day, and weekends too, half the time.

Longboard gathers dust in the shed. I hardly touch my parabolic speakers or my camera.

Surfing and making films is idle time, and I don't have any of that any more.

1996

I specialize in felling big and dangerous trees nobody else will touch. I like the excitement and the skill of learning new things. I help an older guy set up a business doing the tree work for SWEB, and he calls me when they have difficult trees.

It isn't long before I nearly cut myself to pieces with the circular saw.

He's an open saw, fixed to the tractor by a three-point linkage, no guards, nothing, and field muddy as hell. I've been sawing in the same spot for a few days, so it's full of sawdust and I'm thinking, well, if I get off the tractor where I'm supposed to get off him, climbing out of the door, twill be in the mud, but if I get off by clambering out over the top of the saw I'll land in clean sawdust. I think it's a very good idea.

You've got to picture the scene. On the back of the tractor there's an open saw, and there's a PTO shaft, which is a drive that comes from the tractor to the implement to turn the implement. There's no guards, so if anything gets caught it sucks it all up. Well my coat gets caught and it nearly sucks me all up.

I'm stood one foot on the tractor and one foot on the bench, and I'm thinking I won't need my coat, so I take it off, only as I'm shaking off one of the arms he gets caught around the shaft, and he's spinning and I'm still half in my coat and I'm being dragged down on to the top of the bench and I'm right on top of the saw and he's pulling the coat right off my other arm and I only just manage to stop him pulling my arm in with it.

1997

The guy who hired me retires, and his son-in-law takes over, and he knows nothing at all about tree work, and it isn't long before we have a difference of opinion and I walk away.

That's Friday.

Saturday I'm looking in the *Somerset Free Press* and there's a job advertised with the National Park. Chargehand of Rights of Way. Well that's interesting. I don't expect I got much of a chance, being as I got no experience with regards to managing health and safety and writing reports, but I got plenty of experience of Exmoor, so I send off my application and they bring me in for an interview.

Tell us, Mr Hedley, what do you know about Exmoor National Park?

Well, I says, and I don't have to think very hard, it all comes reeling out, all the years of my life I spent topping the fields and poaching trout and deer and following the stags.

I know all the deep oak-wooded valleys and all the paths leading up through them, and where they start and finish and who owns the land. I know the waterways and where she'm likely to flood out. I know where the stags go in spring to drop their horns and where they go in autumn to roar. I know how she gets stogged and soaking wet, like a lake, all poached with the hooves of cattle, and I know how she gets parched and hard underfoot, and I can predict when each of these things is likely to occur. I know she's unconquerable and that's what I love about her.

They'm all kind of looking at me with their mouths hanging wide open and then they start in asking a load of questions like when she was designated a National Park. It was 1954, the year after I was borned, which I know. I do not know that before that she was designated a Less Favoured Area, although I understand it in my bones, because I've had the fortune and the misfortune of being with her and being of her, and carving my life out of her. I know better than they do how she's impossible to farm without subsidies. I know from my reading around the subject that two-thirds of the Park is in Somerset, which is my home county, and the remainder is in Devon. I know there's 40,000 acres of open moorland and 17,000 acres of woodland. I know there's 300 miles of major rivers and streams, and hundreds of miles of smaller tributaries.

Turns out I know more than anyone could have predicted, including myself, and much to my own surprise they call me back and offer me the job, albeit at a reduced salary owing to my lack of formal qualifications.

And that's when everything changes again.

It's my first ever salary, and it includes sick pay and holiday pay. I don't have to go scratching around for work no more, and with a mortgage to pay it's a big weight off my mind.

Job's a doddle, fixing stiles and fences, and strimming paths. I can do that standing on my head. The other part is doing the reports, and even that's easy enough. There are people in the office who think I shouldn't have got the job because I'm not qualified, but there are others who know me and know my skills, and they leave me alone.

Ever since I was a boy pretty much I've always been told I'm no bloody good, and even though I've been in forestry and related occupations twenty-five years and made a success of it, I've been isolated.

It's only now, when I start working for the National Park, and I get my own team, that I start to understand I'm ten times smarter than most of the people above me.

That's how I get my worth.
It's also how I get my downfall.

1997

There's a dark silhouette against the broken grey sky. I'm thinking buzzard, then I hear the angry ek-ek-ek-ek. I stop digging.

I wrap Sizzle in a blanket and lower him gently into the hole I've dug. It's years since I last saw a peregrine, and I take it as an omen.

It's the very start of autumn and it's sort of like the sun is gathering up everything she's got, giving him one last push, starting her long-drawn-out goodbye.

First Sizzle dies, then Lady Di. I hear the news on the radio in the Mini pickup. Numb with shock. Drive home and call up Gillian. She's crying.

We go up together to witness the funeral, the march through the streets, the flowers. Take a day off work and catch the bus out of Dulverton, change at Taunton. It's just the same as when

I went to Germany, only the streets are so full we can't walk even, only totter along a few inches a minute. Never seen so many people, and all of them carrying flowers. We've got our own flowers, which Gillian picked out of her garden. We hand them over to a policeman and he lays them down with the rest by the fence outside the palace. Never seen so many flowers, and everyone weeping and wailing.

I turn to Gillian and I tell her.

You're the love of my life, I says.

I'm aware of the silence. It's so noisy all around, and all I can hear is silence.

Oh Tommy.

I put my hands on her shoulders. You gotta leave him, I says. We can't go on like this forever.

Can't we?

No, I says.

She lifts my hands off her shoulders and gives them back to me.

Don't go giving me ultimatums, she says. I'll be ready when I'm ready.

We watch the procession and I can see Gillian is crying.

Those poor little boys, she says, and them walking along behind their mother's coffin.

Children need their mothers, she says, and I know she isn't going to leave him, I know it then, but I bury the knowledge down below, down in the murk, with all the other things I don't want to know.

It's not until Father's funeral it rears its head again, and that's the last time.

1999

He's in and out of hospital a year or more, and they never do find out what's wrong with him. Then one day he goes in and never comes out.

I've been promoted to Countryside Projects Manager. Pete the depot manager retires and his job comes up and Gillian talks me into putting myself forward. I tell her I don't have a rat's chance, but she goes on and on. She has a higher opinion of me than I got of myself.

Only I'm right. They give the job to someone from the office. Only then he suddenly decides he can't take it and I'm next in line. People in head office don't want to give the job to me, but there are others who know my skills and persuade them. So right from the start there are people in head office who aren't happy.

It's a step up and a big step. I'm running all the field services, forestry, rights of way, and the depot. I've got a team of practical staff and it's keeping me pretty busy.

I go in to see Father in the hospital and I tell him about my promotion, and how my first task as Countryside Projects Manager was to repair a bridge he himself helped to build.

Tell him how I called up Winston Pincham's missus to borrow a Land Rover, and she referred to me as Mr Exmoor National Park. Well that tickles him pink.

He's all shrunken, and tubes coming out of his arms, and I can't hardly hear what's coming out of his mouth. Got to lean in close. We haven't always seen eye to eye, only he seems proud of me now, and I'm holding on to that.

I'm holding on to other things too. Mother and Father walking up to Dulverton cinema from the cottage in Brushford, two miles there, using the shortcuts, and three miles back along the lanes, in the pitch dark. It's the only time I ever saw them happy, except when Father was drunk.

Father was the life and soul of the party when he was drunk. But he didn't get drunk too often. Mother held him in line. Mother wore the trousers. Her was boss. Held us all in line. *Tried* to hold me in line. Father was strict sometimes, if Mother kept on at him, but other times he rebelled. Like the time I had my leg in the calliper and Mother picked me up by my busted leg and dragged me about the place. Father come in and made her stop. There was an almighty bloody row. I slipped out and when I come back every saucepan had a dent in it. Mother had a hell of a temper. Father didn't.

I go along to the funeral on my own, because Gillian and me are still skulking around. I've been on at her about how the kids are old enough and she got to choose, but she can't or won't. Mother's there, tight-lipped as always, but I turn around just before the coffin goes in to be burned and I see her hands twisting in her lap, and a look on her face and it's like the tide has gone out only he'm never coming back in. I think that's what causes me to take the action I do.

Gillian clings to me. I tell her I need more than a part-time wife and she clings on like she's drowning. She weeps and wails, but I stick to my guns.

I've had more'n half a decade of secrecy, I says, and I want a woman I can introduce to my friends. I want someone I can share my life with, in all its parts. And if I can't have that I'll have nothing.

It's a dark stretch of road.

2015

The days without Hope are passing, like all my days, only even slower on account of waiting for her to come home. She calls every night and Catrina puts the phone on speaker and wedges it in my hands – I got to teach her how, make sure I got my hands curled around it properly so it don't fall and clatter on the floorboards.

Hope tells me how she's been to see the beavers and the otters, and I'm glad to hear the smile in her voice. Sometimes I ask her, are you happy? And she says yes I am, and we do get a lot of pleasure out of small things, more'n I thought was possible. We'll pack a picnic and eat it in the garden, listening to the birds, or go to Gilfach and sit in the hide by the river and wait for the dipper.

We'm ever so close, only it occurs to me now how I'm often saying no to her. She'll say let's do this or let's do that and I'll say no. It isn't fair. I got to stop saying no to her.

Sometimes I wish I could pass on, so Hope can have a fuller life, but then I think no, perhaps she don't want that, and perhaps I got to stick around until she's hard enough to cope with the loss. That's my job from now on, to comfort Hope and really try and make her as happy as I'm physically and mentally able.

1999

A new study finds that one square kilometre of Amazon rainforest can contain about 90,790 tonnes of living plants.[17]

2000

Century ends and a new one starts.

I go up and down to Cornwall chasing hurricane swells with Andy Harrison, surfing bigger and bigger waves, scaring myself half to death. I go to visit Mother at the cottage in Brushford every Sunday.

When're you going to get yourself a wife?

I've had plenty of wives, I tell her, only none of them my own.

I get myself another Samoyed, find him in a rescue kennel. He goes by the name of Spike, and oh boy does he cause a stir among the team at the National Park.

He sits on the passenger seat of the Mini pickup with his face out of the window and his ears streaming back, like a girl with a full head of white-blonde hair. He's a picture. People call him Chewbacca and they take the piss out of me because he's all white and fluffy.

You got a bloody pouf's dog. Trust you to buy a pouf's dog. He in't good for bugger all.

You watch, says I, and I says Stags! Stags!

I can feel him tremble, and I says get em boy, and cor and off he goes.

You never seed a dog run so fast in all your life. He'll go down the valley and up the other side.

Ah, matey says. He don't know what he'm doing.

I says you watch, he'll be right on their heels in a minute, and there he is, on t'other side, going through the heather, bloody bounding along, and he's right under, he's right on the ass of him.

Bred for it, I says, chasing reindeer.

I aren't half proud. Only trouble is getting the bugger back. He'll jump out of the pickup while we're going along at thirty mile an hour, land on his bum, take off, and we'll waste the whole bloody morning trying to get him back.

I take him to Exford show and leave him in the pickup, only he manages to get out. Some woman comes down laughing.

Cor, she says. There's a big white dog up there and he's got about ten terriers hanging off him.

Little bastard. Window is broke. If you lean on him he just drops down. God dear there's terrier blood everywhere. They couldn't get into him because his coat was too thick.

I've always enjoyed my own company but now it's hellish lonely and I'm drinking and getting into scrapes.

Tip my van over twice in one night, a Bedford Beagle. I've been up at the farm with Johnny, and I'm filled right up with whisky, and I'm driving home, and I notice the brakes are lopsided, pulling to one side. I think I can cure them by stamping on them hard, so I wait for a straight bit of road and that's what I do. I stamp on the brakes, and they shoot me up the bloody hedge and tip the van over.

Well fuck this. I start to leg it, only then I look behind me and I see the bloody thing there on its side in the middle of the road and I'm thinking somebody's going to crash into it and kill hisself.

So I go back and climb inside it and wind down the passenger window. I jump out, put my back in the van and tip him back up on his wheels again.
    Bloody nuts.

I gather my senses and let him freewheel in a straight line, then I turn the key and away he goes. Only when I get to the bottom of the hill before the lane to the cottage I do the exact same thing again. Bloody stupid. Put the brakes on too hard, tip him over, wind the window down, tip him back up, jump in and drive home.

Then when I get home I slam on the brakes outside the house and the windscreen falls out and smashes on the bonnet. I'm cussing like buggery, but I'm lucky really. Apart from the door handles and the wing mirrors, that's the only damage. Bloody lucky.

Then I get a Triumph GT6, and Johnny gets a Porsche 911, real old vintage cars, and one night we're at a farm called Plymptons, and we decide to race home at a hundred mile an hour.

Well I'm thinking I'd better get out in front of him, so I leave a fraction early and drive as hard as ever I can and I come flying down the road and round the bend and I fly right out over the side of the road and into the ditch where people tip their rubbish, and there's cookers and old fridges and all sorts chucked in there.

Well I'm all jammed in and still struggling to get out of the car when I hear Johnny go past in his Porsche, hunting me down. I'm expecting him to see me, but he never does.

There isn't much I can do except wait, and sure enough a lorry comes along and the driver sees my car there, all jammed in among the fridges and the cookers and he stops, jumps out and walks back.

What the bloody hell are you doing down there? he says, I never thought there was a car park there.

Well, I says, I'm stuck.

He helps pull me out, and it's a remarkable thing, I never done any damage to the car at all, hardly. I jump back in and drive back up to the farm, and Johnny's there and he's made a cup of tea and he's sat down drinking his tea.

Where the bloody hell have you been? he says. You just disappeared off the face of the earth. I thought you must have gone on home.

No, I says, I only ended up in the bloody tip.

Gillian calls me sometimes, and sometimes I call her. We cry down the phone. She tries to get me to change my mind, but I can't. I've made a choice and I got to stick by it.

You love someone that intensely you always think about them. It's like we were joined. I had to cut myself free, but it took the bloody skin off.

2015

Get him will you? Wyn!

There's mud on the lane and I don't want to get my wheels in it and run it all over the kitchen.

There's a great heap of manure piled up on the side of the lane and I can see liquid draining out and heading straight for the river.

Puts me in mind of C. J. White and Partners, of Shircombe Farm in Brompton Regis. Polluted a two-and-a-half-mile stretch of the river with slurry. Had to pay £18,000 to clear it up. Don't see them handing out fines like that now. Bloody shit.

When I was a boy nobody ever planted until the moon was growing. These days farmers wouldn't see the moon if it was hanging off the end of their nose. Too busy spreading God knows what all over the fields, not thinking where he'm going to end up.

3

Fertilizer consumption has jumped from 40 million to 280 million tonnes per year since 1950.[18]

2002

The forestry gang is still in my charge, and I got to report there every Friday afternoon to see what's going on. I've given up all thought of meeting a woman and I hardly notice there's a new face in the woodlands department.

She's got curly red hair cut short and tucked behind her ears, real tomboy. She's got a faceful of freckles, and I never seen her dressed in anything other than jeans or old corduroys. The other girls make an effort with their make-up and the office girls wear skirts and all that, but I never see Hope in anything but old jeans. It's true she'm working outdoors, which is generally the male preserve, with most of the women in the office.

It falls to Hope to accompany me across the moor when I go over to meet up with a farmer who's renting the land on the cliffs above Lynton and Lynmouth. She sits in the passenger seat with Spike on her lap, and I point out all the things I love. It's late spring and foxgloves are out and cuckoo is calling.

It's a lovely drive over the top of the moor. I go up past Dunkery Beacon and down past Horner Wood, and through Porlock and out towards Lynmouth. We take the toll road past Lynton, where you can walk along to the valley of rocks, and go along right up behind Lynton, and I'm pointing out all the landmarks and noticing how Hope seems to be enjoying herself, and she's good company.

I ease the pickup off the tarmac and onto the farm track and Spike puts his paws up on the window and sticks his head out and Hope's laughing and the sun is shining and the sky is bluest blue and little white fluffy clouds skidding across.

The farmer's up there already.
  Watch your bloody dog don't chase the goats, he says.
  Oh, I says, I'm not worried about goats. He'll chase a stag off the side of the moon, but he'm not interested in goats.
  Truth is, Spike's never seen a goat before, not up close. It's me decides goats aren't worth chasing, and bugger me I'm wrong.

He takes off like a bloody rocket. I watch him tear over the top of this scree slope about fifty yard up ahead and farmer's standing there open-mouthed.
  Should have put him on a lead, I says, and I'm proud. He'm hellish fast.
  Still hasn't occurred to me what farmer is on. He's got to spell it out.

You should catch him, he says, before he goes off the side of the Valley of Rocks.

That's when alarm bells start ringing and I start to understand the trouble we're in.

The cliffs overlooking the Valley of Rocks are a hundred feet high, with boulders the size of tables massed below.

If tide's out he'll hit the rocks. If tide's in, waves'll be crashing against the base of the cliff, and Spike's a terrible swimmer. All my Samoyeds were. Too much hair. Drags them down.

I tear off after him, yelling his name at the top of my voice. Spike! Spike! I run to the top of the slope, and when I get there I think fuck. He'm gone out over.

I must've been running bloody fast, because the other two are miles behind. I walk over to the edge of the cliff and look down but I can't see nothing except rocks, the tide's out. Then I see a trail of white hair going off over the side of the cliff and I'm thinking, I'm going to have another bloody heart attack.

There's a gentle slope at first, and I scramble down as far as I dare, thinking if I slip I won't be able to stop myself going over. I find more white hair as I go on down.

When I get back up top I find the others waiting for me, and they'm worrying I'm gone over. I ask the farmer if there's a path, and he says yeah. Tis bloody dodgy, he says, but I don't care. I can hardly hear him my heart's beating so loud.

He leads me along a few hundred feet and shows me a goat path winding down through the slippery scree. It's hellish steep. The last thirty feet is a straight climb, and I've been scared to death of heights ever since my outing up Crib Goch, but I make it. I clamber along the boulders, looking upwards, trying to spot the pinnacle where Spike and the goat went over. I see the goat first. He's splattered on the rocks, blood everywhere, spleen punctured, dead. Christ. I sit myself down on a rock and prepare for the worst. Then I see Spike, and he's just sitting up, looking out to sea.

Spike!

He turns and starts ambling towards me, like we're in the bloody garden, only he's bleeding from the back end.

Stop! I shout. Stay!

I run towards him, hopping from boulder to boulder, and I gather him up in my arms.

He's a heavy dog and it takes me quite a while to pick my way over the rocks to Lynton.

Hope's driven the pickup down, on the farmer's instructions, and she's waiting in the car park. Spike sits on my lap and Hope drives us back across the moor and straight to the vets in Dulverton.

That's some bloody lucky dog, says Mr Peters. Don't seem to be no internal damage. The bleeding is caused by a melanoma and we can take that out. All he done is knock a tooth out.

I'm still worried.

He'm not himself, I says. He'm awful quiet.

That's shock, says Mr Peters, and truth is he might still die of it. We'll sedate him, and if you stay up with him all night then maybe he'll stand a chance.

I spend all night stretched out on the floor next to Spike, talking to him and stroking him, and in the morning he's right as rain, and something has changed in me, too.

It's a Friday. I call up the office to explain I'm not coming in to work.

That evening Hope rings up to see how we're getting on. Spike'd like to see you, I says. He wants to say thank you. She comes to the cottage with a bottle of gin and a packet of dog biscuits and we sit up half the night talking.

She tells me about Wales and how she'm homesick for the places she knows up there and her little cottage. She tells me about her past lovers and I can feel my hands balling up into fists the way they treated her, and she'm a bloody stranger.

I tell her things too.

I tell her how Gracie went to find us a farm only she never came back and ever since I've been dreaming of it, only now instead of a farm it's a piece of woodland I'm dreaming of, a piece of oak woodland. I've been thinking it through for years, how I'll thin the oak and manage it for cleft oak stakes, which I'll sell to the farmers for fence posts. I'll make sure I've got enough firewood and sell what I don't need. It'll be my retirement. I explain how I've already got my winch and my topper, from my forestry days. How I'll need a tractor, and something for storage, like a shipping container.

I've got it all planned out.

I'll build a little hut and plant trees all around. I'll have downy birch and silver birch and holly and cherry and hawthorn. I'll have hazel for thatching spars, and crab apple for the blossom. I'll have oak and ash and Scots pine, trees that'll take longer than my lifetime to grow, and they'll be my legacy. I've been saving money, I tell her, only I can't save it as fast as the price of land is going up.

The estates are buying it for their pheasant shoots, and they'm killing everything that gets in the way of that. Developers are buying it thinking they'm going to be able to make a bundle of money by building a load of cheap houses. Locals are being shoved off, and animals too.

During the First and Second World Wars the woods on Exmoor were cut for timber. Then afterwards, in the 1960s, they took out the remaining broadleaves and planted with conifers. Douglas fir, Sitka spruce, Japanese larch, western hemlock.

Hope says it's the same thing in Wales. One day there's a forest and the next day a desert of stumps, and trees planted so close together nothing can live in between them, and even the birds stay away.

2009

It's coming up time to leave the hospital and I don't know what's going to happen.

How am I going to cope?

I can't do a single thing for myself.

Who's going to do my bowels?

Who's going to empty my bag?

Don't let your wife do it, say the nurses.

She'm not my wife, I tell them, but I know what they're on. That kind of thing will murder a relationship.

I've got constant stomach aches from not being able to digest the food I eat. Kidney's fucked, I expect. That's what happened to Father. Put him on steroids and it ate the fucken lot.

I'm glad Father's dead. Would've broken his heart to see me in this condition. He'd have died from the news of it.

Everyone keeps on saying try acupuncture, try cranial, try reflexology, everyone and their bloody farm dog. I'm fed up with it. Just bugger off and leave me alone, I tell them. And most of them do in the end.

It's Hope drags me through.

I wake up and she's painted my toenails, each one a different colour.

You're lucky to have her, says the nurse.

Oh, I says. Hope's not luck. She'm a miracle.

2003

Some people go out looking, but I've always been a lazy bastard when it comes to relationships. You gotta almost jump on top of me and rip my trousers off. Hope is different. From the day Spike jumped off the Valley of Rocks I'm fixed on it. She'm the one.

I take my time, take it slowly, and we'm just friends at first, rub along and have little trips out together. This goes on for a couple or three months. She'll ask me round for supper to meet some of her friends, and then I'll trundle home. I'll invite her to the Poltimore to meet Johnny and Horatio and the others. Then she invites me up to Wales for Christmas.

Problem is, I've already promised Heather and the boys I'll spend it with them.
    Well, I says, I'll do both.
    So I eat my dinner in Dulverton with Heather and the boys, then I set off, and five hours later I eat another dinner up in Wales with Hope and her parents on their smallholding.

It's late when Hope leads me down through the woods to the cabin where I'm sleeping. There's a dusting of snow on the ground and thick stars above and the cabin is hidden halfway down the side of a valley, in amongst a load of old oak that escaped the felling because of the steepness of the slope. Tis bloody lovely. She's been down and lit the fire, so it's toasty warm, and there's no electricity, so we have to light a load of candles.

That's the night we get close to one another, and I'm just so gobsmacked. How it could all come together after all that time and all those decades on my own.

Takes me fifty years to find her, and I've been fending off the day I lose her ever since.

2004

We wake up together in the cabin.

I leave Hope sleeping and light the fire, then she gets up and makes porridge for breakfast on the old gas hob, which is just like the one I've got in my cottage. It's so peaceful.

I could sit in here all day, I says, but Hope's restless.

I want to show you something.

We walk to a piece of land about a mile away from the cabin, on the other side of the valley. Piece of open pasture, leading up to steeply sloping woodland, thick with oak, twenty or thirty years old, and above that a flat area, buried in bramble.

It's for sale, says Hope.

Well, I says, it's perfect.

Thirty grand, she says. I checked.

Well, I says, and I'm looking around me thinking how this is what I've been dreaming of all this time. Hope is watching my face.

Problem is, I says, tis two hundred miles from my cottage. Twould be hell of a commute.

2004

The first global assessment of soil loss, based on studies of hundreds of experts, finds that 38 per cent, or nearly 1.4 billion acres, of currently used agricultural land has been degraded.[19]

2004

Hope is at the stove, cooking one of her vegetarian meals.

Tis silly to be renting a cottage of your own, I says, when you're spending all your time up here.

Reminds me of Heather all those years earlier. I wonder if Hope is feeling the same way I did.

It can be our little home, I tell her, pleading.

I can see now what I couldn't see when Heather was pleading with me to live with her. I wasn't in love with Heather. I wasn't in love with Gillian either. I'd never been in love with anyone before I met Hope.

I talk her round and the truth is she's easily persuadable. We're at the point when we don't want to let the other one out of our sight.

The adjustment is easier than I'm expecting, after twenty years of living in the cottage on my own. Only thing I find hard is the way she uses all my pots and pans. I've kept them nice, and they're nice to cook with, and then Hope comes along and scrapes them all up and leaves them in the sink to go rusty.

I go to visit Mother in the cottage in Brushford every other Sunday as is my habit, and then one Sunday in August I take Hope along.

When we're leaving I pull Mother to one side.

If this don't work out, I says, I'm not going to bother. I'll become a weird old man, just collecting antlers.

I get myself an old Iveco van and kit it out with a bed and a cooker and we go off for the weekend, only half an hour later we've turned around and we'm coming back again, just because of something she said.

It's the first time I've ever really shared my life with anyone, not holding back, and it turns out I've got a hell of a temper. I'm always the stroppy one. I'm always the one to say fuck off I'm going home. I'm insecure.

I can't understand what she sees in me, and I cover it up by being wild and argumentative.

You got to learn to shut up on certain things, says Heather, when I go storming down to hers one evening.

From time to time my mind returns to the patch of land we went to visit on New Year's Day. I picture myself knocking in fences and planting trees.

It takes my mind off work, and how things are turning sour at the National Park.

2004

Christmas comes round again, and we've got a load of cash in a drawer at the depot, from selling logs on the quiet. The money is for spending on a slap-up Christmas meal, and everyone's in on it, including the office people. They don't want to fork out on a Christmas party for the workers. It's my job to organize everything and to put in a bit extra if we'm short.

We've got a bottle of Laphroaig in the office, which is a gift from Eagle Plant. I pick it up and take it out into the yard, where the boys are gathering. Hope's there with the woodland team, and Alan Hollis is there from the depot.

Alan, I says.

We aren't best of friends – he never thought I ought to have been promoted – but we are civil.

Tommy, he says.

I open the bottle of whisky and pass it around, and Alan Hollis drinks more than his fair share, which doesn't surprise me. He's that type of man.

Then the taxis arrive, and I stash the bottle behind the pallets so we can finish it off later. There's only enough money for each person on the team to have two drinks with their meal, and one afterwards at the bar.

The meal goes off well, and I'm standing at the bar afterwards with Hope and some of the boys, feeling pretty good, happy, when Mary Hollis comes up behind me and pinches me on the bum, right in front of her husband and with Hope standing there too.

She smiles like we've got some sort of prior agreement and she's got teeth like a row of condemned shithouses, reminds me of Mrs Evans, and she's drunk. I push her away. I'm fuming.
 I reach for Hope and sling my arm around her neck. I pull her with me towards the bar.
 Who's that?
 Mary Hollis, I says. She's a witch, and a fat one. She'd need two fucking broomsticks to get up the hill.
 Ssh, says Hope, but I'm still fuming.

I'm fuming even more when I get back to the yard with the team and we find the empty bottle of Laphroaig and Alan passed out behind the pallets. Greedy little bastard. Well I wake him up and load him into a taxi and I get in after him so's I can unload him at the other end, and her's just standing there in the driveway and I do not mince my words. Them boys were coming back, I says, and he'm drunk all their whisky.

I know that from that night onwards my days at Exmoor National Park are numbered. When we go back after Christmas there's a group saying they'm unhappy with me as boss. I push them too hard, they're saying, and they'm the laziest bunch of bastards in the place.

I expect people to work hard and take pride in their work and we achieve a lot, I tell the bosses when they haul me in.
    Alan's among them.
    But some of the boys don't want to do nothing but shoddy work, I says. All they want to do is hang around until they get promoted to the office and can sit in the warm behind their desks all day, gossiping and drinking coffee.
    I lock eyes with Alan.
    Don't want to get their hands dirty, I says. Well they don't like me, and I can be rude, I'll accept that, but it's my job to make them get their hands dirty.

It isn't just the team. It's the whole thing. Going in the wrong direction. It's harder to get anything going, get anything done.

We used to have projects like the sawmill at Simmonsbath, and the ditches I designed and dug with the team, to bring back the peat bogs. I've been working the park ten years and I've learned to really use my head. Only now it's more like having to use my cunning, like a fox.

I'm dreaming of just sitting in a digger like I used to, with no team below me and no management above me, doing a perfect job with my eyes closed.

2005

Severe drought in the Amazon.

2005

I'm not altogether surprised when Hope sits me down and tells me she's thinking of applying for a job back up in Wales, with the Woodland Trust.

Hope is one of those people with friends in every country and every corner of every country. She's got friends everywhere and they'm always sending her notices for jobs. They all want her to live close to them, which I can understand, although it does make me insecure. I try not to show it. I love having her with me up at my cottage, but as time goes on I can tell she's hellish homesick.

Well, I says, I'm not going to hold you back in your career. If you get the job and you move back, we'll just have to divide our time.

2006

It starts with rumours, some of the lads are unhappy, then one evening a young boy called Eddie comes round to the cottage with a petition some of the team have got up against me. Complaining they don't get enough breaks throughout the day, that sort of thing.

He wants me to know first-hand he hasn't signed it, and neither have any of my core workers. But plenty of others have. I glance at the names then crumple it up and throw it into the corner of the room. I never felt so betrayed in my life.

Hope says I should hold on, wait it out. Three years and I can retire on full pension of seventeen grand a year, she says.
    But the office says they want me inside, writing reports.

My heart is aching to buggery, wants to break right out of my chest.

The headquarters of Exmoor National Park used to be a workhouse. I go in through the main doors and up to head office. I tell them they can stick it. Then I walk back out of the main doors and along the street and out and down to the river.

2

The number of dammed rivers has risen from 4,000 to 28,000 since 1950.[20]

2010

Doors open automatically. My hands and arms aren't strong enough to turn the wheels on the chair, so Hope has to push me through. I've got some muscles left in my arms, but most of them are gone – it depends if the nerve endings are above or below the break. I don't recognize my hands. They're twisted at the wrists and the skin is stretched like they've been burnt in a fire. You look at my arms you can see the bones under the skin, bones poking out under the tattoos.

We go along the side of the canal. I'm trying to focus on the trees, how they look like bony fingers poking at the sky. It's winter and everything's dead or dying, even the grass. I'm trying not to look down. The ground is moving too fast. The water in the canal is black and the towpath is crowded with people walking and running and cycling. Hope pushes me right up to the edge, and she's flying along with the chair going in front of her, and she's so happy I don't want to tell her I'm frightened, how all I can think is how I'm going to fall out into the water and drown.

2006

I swing back to forestry again, only I'm leaning more and more towards the conservation side. I get work with the RSPB, Woodland Trust, English Nature. I even pick up contracts with Exmoor National Park, sidestepping head office. I get paid twice what the contractors got when I was manager, and I don't have a bloody depot to worry about, neither. I got my quad bike, my trailer, my two chainsaws and the hand winches, and that's it.

I like the conservation work. It's challenging in a good way and I can take my time over it. Timber production isn't the main importance. I got to put a lot of thought into it, balancing the different parts. Letting light in, encouraging wildlife, encouraging the remnants of ancient woodland to grow back and flourish.

When I can't get restoration work I do general forestry.
    Contract felling, like I done before, or I buy standing timber and fell it and sell it commercially. I'll look at a job and I know nobody'll touch it. I'll scratch my head.

Two hundred pounds a day, I says.
Oh, that's a stiff one.
Take it or leave it.
Well they always take it.

The year after I leave the National Park I earn fifty grand, and I get plenty of time off. I'll price up a job and finish it two weeks early and I'll have already allocated myself a week's rest between contracts, so that'll be three weeks.

I spend most of my time off up in Wales with Hope. She's staying in her parents' house while they'm off gallivanting somewhere and there's a heap of work looking after the birds and the horses and the sheep.

Then I go and destroy myself again.

The job is a PAWS – Plantations on Ancient Woodland Sites – and I'm working with a couple of foresters from North Wales, only they don't want to work.

One drop of rain they get back in the truck. Two drops they go on home, and this is north bloody Wales.

Put one tank of fuel in the chainsaw and when he'm done they go on home.

Well I'm staying up there in my van and I'm on site every morning at eight o'clock and I've done a tank by the time they turn up and we'm all sitting around drinking coffee.

You're working too bloody hard, they tell me, you're showing us up.

One of them, Steve, he's a big chap. Six foot six and carrying a fair bit of weight.

Well, I says, we get it done we get to go home and not come back.

Takes a couple of days but soon we'm all seeing eye to eye and they'm a funny pair when it comes right on down to it. Got me bloody chortling all day long.

The light is fading when we come to the last tree, and we been there three days, and maybe I cut it badly, he's a steep slope. Any road he comes down dreadful, all across the track, with huge branches sticking up in the air.

Me and Steve set to it with the saws, cutting off the tops. I'm standing on a pile of logs, saw wood logs, not firewood, eight feet long, so I'm thinking it'll be safe enough. Only I reach up with the saw and cut off a big branch, and the branch falls down and hits the end of the pile of logs and catapults me into the air, tips him backwards and breaks my bloody ankle.

Got to leave the van up in Llandudno. Steve takes me to the hospital, waits along with me, then drops me off at Hope's parents' house. I hobble in with my crutches, and she'm saying, what you gone and done now?

Oh, I says, just used up another one of my nine lives. Don't reckon I've got many left now.

Bloody right, says Steve, he'm lucky the bloody tree didn't cut him in half.

When I can walk we hobble up to the patch of land and have a picnic. He's still for sale, and with Hope back in Wales I get to wondering what if I moved on up, bought the bit of land, put my dreams in motion.

Only it would mean selling my cottage and leaving Exmoor and I got a hell of a bond with that place.

I turn it over in my mind and it gets so I can't sleep. I don't know what to do.

Then fate steps in in the form of Mr Smythe, and I decide I might as well go on along with him and his plans, since I hant got no better plans of my own.

Not that it's an easy decision. Nearly bloody kills me in fact.

2007

I drive out early to Saunton with my kayak on the roof of the van. It's a red-hot summer's day and the sea's flat as a witch's tit, not a ripple on it. I wade out into the shallows with the kayak and push off. I'm only planning to paddle around the point to Croyde and back again, but I quickly get into a rhythm, and I'm soon stroking towards the horizon, trying not to make a sound when my paddle hits the water.

I can see Lundy in the distance, but I hardly see it. I'm thinking about Mr Smythe. He's one of those wealthy gits who's moved in from the city, I can't remember which city, and he's exceptionally well off, and he's bought the manor house for well over the asking price, and I'm the ruffian neighbour, and he wants me gone. Long story short, he's offered me a rake of money for my cottage.

I could pay off the mortgage, buy a cottage in Wales near Hope, and still have enough left over to buy the piece of land and a tractor and anything else I might need. I can see it all in my mind. I can see it easily, like it's already happened.

The wind picks up when I leave the shelter of the bay, and before long every stroke's a brace stroke, but I keep on going. I keep on paddling.

The sea's turned choppy, and the kayak's tilting all over the place, but Lundy's so close I don't feel able to turn around. I know it's something between eight and twelve miles each way, from Saunton to Lundy, which is one hell of a long way to paddle, especially in choppy seas. I know I'll have to turn around and paddle back at some point. I got no food, and no water either, but I keep on going.

I get within three-quarters of a mile, close enough to see the day trippers wandering around on the headland, before I finally turn around.

It's even harder going back. I'm tired from the paddle outwards, and weak from not having any refreshments. It's coming on late afternoon by the time I feel the kayak hit the sand. I try to stand up, but my legs give way and I fall backwards into the water. My right hand's all seized up from gripping the paddle.

I use up the last of my strength dragging the kayak up the beach. I go back to the van and change out of my wetsuit and into my clothes, then I go to the shop and buy a Mars Bar. I allow myself a Mars Bar.

I wait for Andy on the tables outside the King's Arms. I've already had two pints by the time he arrives and I'm halfway down my third. The sun is glinting off the roofs and the young herring gulls are crying and wailing and I'm thinking of all the times I've had thereabouts, all my life up there on the moor, and my newer life out there on the coast. I've got some wonderful memories along with the bad ones, and the good ones far outweigh the bad ones. I got to shut my eyes and blinker to it until I reach a point where I can't go back.

I take Mr Smythe up on his offer on the condition he keeps the cottage as it is. Should be listed, I tell him, as cottage was the lodge and the manor house is listed, but somehow he's been overlooked.

I won't never touch it, says Mr Smythe.

Well I'm hardly out of the county before he tears it down.

2007

The urban population exceeds the rural population for the first time in human history.

2008

I buy a little cottage in Wales, about five miles away from Hope's cottage, and I pay a hundred and forty grand for that, and then I buy the piece of land, and I pay thirty grand for that, and lo and behold I've reached a point where I'm secure. I own my own property outright and I own some land. I'm in a good position.

It's a hell of a wrench.

I still wake up some days not knowing where I am. I wake up in a panic, in a hot sweat.

I force myself to put a mental block on it.
    I got my woodland. I got everything I ever bloody wanted. I tell myself how lucky I am and it's true. I'm very, very lucky.

I don't know the glass is about to tip over and all my luck's going to run right on out.

2008

A team led by Professor Tim Lenton[21] introduces the term 'tipping element' to describe large-scale components of the Earth system that contain tipping points, or critical thresholds, where a tiny perturbation can qualitatively alter their state or development.

The term 'tipping point' feeds into 'the popular notion that at a particular moment in time, a small change can have large, long-term consequences for a system, i.e., "little things can make a big difference"'.

The team identifies nine potential policy-relevant future tipping elements in the climate system:

Arctic sea-ice
Greenland ice sheet
West Antarctic ice sheet
Atlantic thermohaline circulation
El Niño – Southern Oscillation
Indian summer monsoon
Sahara/Sahel and West African monsoon
Boreal forest
Amazon rainforest[22]

2008

I put in a shipping container and get my tools up there and my little tractor. I plan it all out carefully, what I'll put in, and we start planting. It takes us all winter, right up until spring.

I buy a load of nest boxes for the birds and hammer them onto the existing oak.

I go and pick up Wyn from a farm up in Yorkshire. I thought it would be nice for Spike to have a bit of company, and with him getting on. I need a dog. Don't know where I'll be when Spike dies.

There's only one more thing left to do, and I wait for the right moment.

I go down on my knee.

Hope, I says. Will you marry me?

2017

What did she say?
   What?
   When you went down on your knee?
   She said she'd think on it.
   How did you feel about that?
   Oh I felt fine. I was confident she was going to come round. I didn't know it was all about to go multi-pear-shaped.

Catrina's halfway in the hedge and she'm only wearing shorts.
   How's the blackberries?
   Not as good as home.
   Devil spits on em in October.
   It's still September.
   Well that's a relief.
   Are you scared?
   Yes.
   She'm got blackberry juice all down her face.
   Let's have one.
   She tips a handful of berries in my lap.

You'll have to do better'n that, I says, and I use my right hand to make my left hand into a kind of bowl.

She scoops up the berries and puts them in my left hand, and I can sort of get them in my mouth like that.

Kids used to get one and a half days a week off school to pick blackberries.

Did you?

No, it was before my time.

Catrina's come up early. She'm going to be staying in the cabin in the woods where Hope and I first got close. Hope's cousin wanted to stay there but we'd already said yes to Catrina. They'm all coming. Johnny and his new partner, Keith and Bev, Susan, Andrew Rudd, Reenie, Heather, Hope's brothers and sisters, all their kids and all their cousins. Every single person we invited said yes. It's going to be a full house, like Mother's funeral.

Hope's in London getting her final bits and pieces. Andy Harrison's coming up tomorrow to go over it all and keep me company while Hope's gone, along with Catrina.

Did you bring your cello?

Yep.

What're you going to play?

The one you asked for.

It's very quiet in the lane. Only sounds are the river and the wheelbarrow and Catrina's questions. Never bloody stops. I don't mind. It's nice in a way, to have someone listening. I'm trying to avoid the puddles, because if I don't I'll take a load of mud in the house with me and get it all over the floor.

Catrina's in a fidgety mood, keeps getting in the way of the chair. It's like a dance, me going from one side of the lane to the other and her crossing over and getting in my way.

What actually happened?
How exactly did it go multi-pear-shaped?

It isn't the first time she's asked, and I usually behave like I can't remember too clearly, only it isn't true. I can remember in great detail, like it's happening on a screen in front of my eyes. It's a story I got to read again every single day, the story of my life, and I want the ending to be different, only the ending's always the same. It's like a nightmare, only I never wake up. I still get emotional sometimes. It depends on how much pain I've had, and for how long I've had it.

Sometimes if I've had pain for a fortnight solid then it wears me down and I get to thinking about how the ending could have been different. I could have been fine. It all could have been fine. But I'm stubborn, and I'm stupid, and I overreached.

11 March 2009

I pull out onto the A470. It's still dark. I've got the *Today* programme on the radio. Spike's in the back, and he'm eleven year old now. Wyn's just a puppy. Stop for fuel in Rhayader. It's cold for March, even for Wales. Cold hands. I'm peering in at the dogs through the bulkhead. Spike's sleeping. Wyn's chewing up the seat belt.

It's thirty-five miles through to this little village in the upper Wye Valley. Erwood. Still there. I got to pass it every time we go south, which isn't often nowadays. Spectacular scenery. I do enjoy that. The Aberedw Rocks and the Radnorshire Hills. Reminds me of Exmoor a little bit.

There's a pool of mist lingering in the upper reaches of the valley. I go in underneath it and fancy I'm catching glimpses of the river down below through the trees. I see it in my mind's eye, wide and smooth as glass.

I've been hired by the Woodland Trust to plant a load of trees, and they'm all there waiting, a crowd of saplings in the bottom corner of the field. Field is steep, but I'm used to that. Don't think nothing of it. There's three hundred wooden stakes next to the trees and three crates of plastic rabbit guards. Apart from the dogs, I'm on my own.

I park the van at the bottom of the field near the saplings and let the dogs out. I unload the quad and the rest of the tools from the trailer. I rummage in the cab for my flask and my sandwiches. Eat half the sandwiches for breakfast and put the other half away for lunch. Whistle for the dogs. Spike has started having trouble with his back legs, so I lift him up into the front of the van and settle him in on the passenger seat. Wyn I tie up to the chassis on a long rope, so he won't chase the quad.

I load up the quad with the first half-dozen saplings, stakes and guards. I work all day, navigating the twists and turns and rabbit holes of the steep slope on the quad, same as I've done a thousand times before, in a thousand different woodlands.

Towards the end of the afternoon, I let Wyn off the rope, let him chase the bike up the field. I'm gunning it up the steep slope and I can feel him slipping out of control, rearing up like a horse, and next thing I'm falling backwards and then I'm upside down and the quad's on top of me.

I lie there for a few moments, not long, then I heave the heavy machine off my chest, roll it over, haul myself up to my feet. I rock the quad side to side, trying to get him upright again. I shove it, put my shoulder to it, and that's when I feel my neck crick. I get him upright, then I move my head from side to side, to uncrick my neck. It feels stiff, but that's all. I drive the quad back to the bottom of the field. I check the time. It's four o'clock. I decide to call it a day.

It's dark again when I turn off the A470. I go to Hope's cottage instead of my own and I tell her what's occurred and she says I do look a bit pale, and perhaps I ought to see a doctor.

No, I says, I'll be all right.

Next morning I go back to finish the job. It's raining, and Spike doesn't like the rain, so I leave the dogs behind. I plant the last tree at five, then spend an hour loading my tools and the quad back into the trailer. I had planned to stop for a pint in the Wheelwright Arms, but my clothes are wet, and I'm still feeling a bit groggy, so I decide to go straight home.

Next day is Friday. I load the dogs into the van and drive to Newtown, drop in on some old forestry friends of Hope's, Pete and Kerry.

Kerry opens the door.

What happened to you?

Oh not much. Flipped the quad. Put my neck out.

Well, she says, you don't look too good. You ought to get yourself checked out.

Pete comes in.

What you done?

Cricked me neck, I says. Can't seem to uncrick it.

What about that chiropractor? says Pete.

That's right, I says. I'll make an appointment to see the chiropractor. I've seen him before. He'll know.

I've had ample occasion since to wonder what might have occurred if the chiropractor had seen me that afternoon, but he can't see me until Monday, so I go on home, and the next day I go up to my own woodland with Hope to check on the trees we planted.

It's a bloody lovely day and we're up there right until it gets dark, clearing the flat section up top. The young trees are doing well. We put in a little bit of oak and lots of birch and some thorn and a bit of cherry and quite a bit of hazel because I want to make thatching spars and ribbon hurdles and walking sticks. And I'm going to make sure I get enough firewood, enough for me, and enough for Hope.

I've made a hard standing for the container, and I've got my little tractor down there, and my topper and my winch. I've been working and planting and clearing all winter, every weekend, and I've got it all tidy and nice. Soon I'll start thinning the oak that's already there for firewood and cleft stakes. I'll sell them to farmers for fenceposts and what have you.

I'm fifty-five and I've been working hard all my life, except the years right after my heart attack, and I'm ready to sit back and enjoy it all.

Only I still don't feel right. I'm groggy and I've got no colour, which is unusual for me when I'm healthy. I tell Hope I'll go to the doctor on Monday if I'm not better.

We drop in at the Badger on the way back from the land and I drink two pints of cider. Forget all about my neck. Forget about it right up until I get up to piss at three in the morning. Cider will do that. We're in my cottage and the bedroom's downstairs and the bathroom's upstairs.

I climb the stairs in the dark, feeling my way along the wall with my hand. Don't want to wake up the dogs. I'm half asleep and lazy and I sit down on the loo to piss.

I put my head in my hands and my elbows on my thighs, and my neck clunks.

Crunch.

Well bugger me, I say out loud. Perhaps that'll do it some good. Perhaps I'll feel better now.

I get up and go back along towards the bedroom, only when

I get to the top of the stairs my right leg collapses, gives right out under me, and I go head first tumbling down.

   Hope comes out of the bedroom.

   What happened? Are you all right?

   Yeah, I says. I'm fine. But where's my legs?

   They're here.

   Touch em.

   Well, she says, I am.

   Well, I says, you're not.

   I can't turn my head to look at her.

   I can't feel the floor.

   Shall I call an ambulance?

   My legs are floating.

   Yeah, I says. I think you'd better.

   My legs are gone.

1

Man-made emissions have climbed from about 5 billion tonnes of CO2 per year in 1950 to more than 35 billion tonnes per year today.[23]

2017

The sound of the river and the sound of aspen shaking in the wind, flat stems rattling.

How did you get through it?

I don't know, I says. Inner strength. People think they haven't got it, but they have got it.

Air so still it's like the world holding his breath.

Do you think everyone's got it?

Catrina's staring at me like I'm not there.

Everyone's got it, I says. In varying degrees.

And I had Hope.

PART TWO
ADAPTATION

But it isn't absence that causes sorrow. It is affection and love. Without affection, without love, such absences would cause us no pain. For this reason, even the pain caused by absence is, in the end, something good and even beautiful, because it feeds on that which gives meaning to life.

Carlo Rovelli, *The Order of Time*

2010

Hope picks me up in a little red Doblo. It's all kitted out for a wheelchair. She's so proud to be able to pick me up in her own vehicle.

I sit in the back on my own, my wheels clipped to the floor. I try to get her to slow down, but she can't hear me. I'm worried the catheter tube will fall out and the urine will come out of my penis.

Two carers come every morning to get me up and running. It's hard work, and I never know from one day to the next what's going to happen. The catheter clogs and my stomach goes into spasms. Bowel movements like sitting on a bucket of ice.

We're staying with Hope's parents, because they've got a conservatory on the ground floor. I sit in that conservatory for a month. I've got two electric heaters, hot-water bottles, blankets over my knees. I'm cold all the time.

I'm thinking, oh well, that's my legs gone, I can cope with that. I'll carve things out of wood. I'll play the bagpipes. Only I hant got enough puff for the bagpipes, and I hant got enough pressure in my arms to squeeze the bellows, and I can't hold the chamber, and I can't cover the notes.

And I hant got my hands.

There's a bird table and I watch the birds and the other wildlife, squirrels and rabbits, and the occasional fox come along to steal the chickens. I count siskins and goldfinches, coal tits and redpolls, nuthatches, sparrers, robins, wrens. I wave my arms at the magpies and the sparryhawks, but they don't take no notice. They know I can't do anything. There's a blue tit hammering on the window all day long, picking a fight with his own reflection.

I eat my food from a tray on my lap. Hope sits with me and we watch films. I try to keep it together, for her sake.

I watch the dusk and I watch the darkness. Night is the worst. I've got a bleeper I can press, and it'll sound in Hope's bedroom, like one of those alarms mothers have for when their babies cry. I never press it. I learn to wriggle across the bed, takes a lot of effort. If there's a pillow in the way I can feel the break.

I have to learn everything all over again, and sometimes I get to thinking how it's all for nothing, but Hope finds ways to help me adapt. She finds mugs with handles big enough for me to get my hands through, hands like trowels. She puts strings on door handles so I can open and close them, loops cable ties on my zips so I can tug them up and down.

Wyn doesn't know what to make of it. Yaps at the chair and the stranger sitting in it. Doesn't know who I am, or where his master's gone. If ever he hears a quad bike he'll go berserk. I miss Spike something rotten.

It's February when they deliver the electric chair. March when I leave the yard for the first time, almost a year to the day since the accident. The sun's out and there are daffodils lining the verge. I don't tell anyone where I'm going. I'm hoping for a bit of independence.

I've got a kind of gearstick to adjust the angle of the chair. I'm still getting used to it. I raise it up and tip it back depending on the angle of the slope. I start down the track. It's muddy, but I press on. It's overwhelming being outside on my own.

I can see Hope's mother coming in the van, looking for me. I hide in a gateway. There's a big pile of plastic in the corner of the field, which farmers must've ripped off the early potatoes.

It's worse for Hope, because she doesn't like to see me in pain and she worries. She's worried now. The last few weeks have been awful.

We have to think about the future, she says, when I get my breath back.

I sell my cottage and pool my money with Hope. We get an architect to design an extension on hers. Cottage is tiny and the stairs are narrow and the doors are narrow and it's a big job making it suitable for me in my new state.

It's our joint project, our shared project, and it keeps us going. I trundle over every day to check on the builders, run my chair up and down the lanes, can't go over the common.

Then one day I go round and the builders have cut the holly down after I told them not to.

I told you not to cut en down, I says.
    It was in the way, says the younger one, and he's smirking and glancing at his boss like it isn't just my legs have gone.
    I told you not to cut en down, I says to the older one, and you cut en down. What've you got to say for yourself?
    It was in the way, he says.
    Don't treat me like a fucken idiot.
    Well don't treat me like a child.

I've had enough of being insulted, I says. You can pack up your things and go on home.

I march off up the hill in my wheelbarrow, and when I come back down Hope is there.

Where are the builders?

I sacked em, I says. I spent a hell of a lot of my life being insulted by people, and I've had enough.

What are we going to do now?

Hire some more.

It isn't long before I'm ashamed of myself for sacking the builders. Hope is tired out and I'm up and down like a yo-yo. I can't control it. The previous week I sacked another carer, so now we're down to one, which means more work for Hope. I spin the chair around so I don't have to look at her face.

I'll sort it out, I says, and I do.

I find other builders eventually and they're somewhat more respectful than the original pair.

I'm still fighting.

There's an elemental part of me thinking one day I'll get on top of it and be back where I was.

It takes a long time for it to sink in.

2010

Severe drought in the Amazon.

2011

We've got underfloor heating, so I don't drop dead from hypothermia – my body can't regulate temperature no more – and we've got a lift, so I can go between the bedroom and the kitchen without having to go out into the rain or the dark. I've got my own workshop, with a ramp for the chair and all my tools set out on a bench that's just the right height for the wheelbarrow.

We've got the eight pairs of horns I saved when I left Exmoor, and my pencil drawings of Winsford Hill done by Miss Dorothy, who lived in the big house on the way to Edbrook cottages, and the sofa Winston Churchill once sat on, which I won't be sitting on again.

We've got one room we're calling the study, and we store my boxes of photographs and books in there. The study is for me, principally. Hope's going to work in the part of the house I can't access, which is all of the old part. I've got a table with space underneath for the wheelbarrow and I've got my laptop, which I learn to operate using the end of a blunt pencil. I check my emails every three months, then delete them all.

Hope lays out my trophies, and the framed picture of me in my hunting gear, which was taken not long after Gracie left, when I was in my fightable stage. She finds the one of me on my skateboard in Sennen Cove, and puts that out too. I leave them out for her sake.

It's an ongoing process, learning how to get around, how to adapt, thinking how am I going to open that door, and how will I close it again, and if I close it how will I be able to open it again? I try to learn new tricks, thinking to show people how well I'm coping. I've got a walking stick propped against the corner of the wall by the wood burner and I learn to use it to pick up my hat from the floor and place it on my lap. I use the stick to open and close the vent on the wood burner, only I can't lay the wood burner or light it. I used to love being on my own, but now I got to have people come round all the time and most of them never stop talking.

I put Wyn on the lead and tie him to the chair and we go along like that, and I'm thinking it'll be a good way of getting him out of the way if a tractor comes along, or a quad bike, only then I try it and I reckon it's more likely he'll pull me into the road and kill the pair of us. He shits and then flicks it into my face and there's nothing I can do to get out of the way.

There are so many things I can't do.

I try to go up to the top of the hill, where the farmer's put an old railway carriage for shelter, only I can't get the chair to go over the drainage channels in the lane. Wyn bites my feet as I'm going along and when he doesn't want to go he'll come and stand in front of me and bark and when I tell him to give over he bites my ankles.

Hope gives up her job so she can stay with me, because I can't be left for any length of time. She gets a job she can do from home, although I know she misses being outside.

We organize my workshop. I show Hope the Perazone bottle I hung off the top of a Corsican pine that was growing near Edbrook cottages. I climbed up and hung the bottle on the end of the leader, and he put out a new leader and grew another thirty foot. Then thirty years later I was hired to cut him down, and I found the bottle still there, and I took him down and brought him home.

We hang up my snow shoes and my rabbit traps, and arrange my screws and bits and bolts in their cut-off plastic containers. Hope fixes my power tools onto my workbench, which is just the right height for the chair, fastens them down so they can't move, but I still can't operate them. I haven't got the strength.

I try the sit-on mower, but I just cave in on myself, like a sack of water.

2012

We go down to see Mother in the home in Taunton. Takes one look at me in the chair.

When are you going to get rid of that thing?

Never, I'm afraid, Mother, that's it. I'm skidding to a halt.

Her's dosed up. Doesn't take it on.

My cousin's wife come up to me at Mother's funeral, gets hold of my arm, nearly pulls it off. Makes me feel this hole in her head the size of a tomato. Brain haemorrhage.

Then some woman I don't recognize comes up to me, grabs me by the arm.

We told her not to beat you so much, she says. Leave the boy alone, we told her. You'm always beating him. He don't need it. She was lucky we never called social services. We was ready. There was a whole group of us ready to report her.

Hope is curious.

Mother was most unhappy, I tell her. Even back when we lived in Winsford. She had an unruly kid. Me.

Did she beat you?

I've never told Hope or anyone about the flippy stick and the buckle strap and being shut in the cupboard downstairs with the spiders.

I got water in my eyes.

It wasn't just me, I says. Mother was on at Father the whole time. I think now it was due to unhappiness on her part. Full of nerves she was. When she went through the menopause it was dreadful. Bloody nightmare. How Father didn't beat her over the head I don't know.

I try to get Hope to see the funny side, but I'm wondering if there was a funny side. My eyes are spilling over. I lift my arm and wipe them with my sleeve, arm like a length of cut timber.

All I ever wanted was to please her. Like a loyal sheepdog who gets kicked every day and still comes back.

And now Mother is dead.

2012

I've got a care package, with money to pay my carers, and I've got some hours left over, so I ask a local chap called Miles to come on Friday mornings to help split logs and stack them and do all the little bits of fixing and plumbing and maintenance I'd normally do myself. It's hard work, watching him struggle, but I've given up trying to tell people how to do things the way I'd do them, the way they ought to be done. I teach him how to fix the chainsaws and the lawnmower and the water pump and he'm good company. Gets so's I look forward to him coming.

Hope's got a friend who's got a son who's into skating, and he comes round and spies that old teardrop and I can tell he wants it.

Take it, I says to him, help yourself.

I wish I hant.

I expect him to take it and have fun with it, only Hope's friend comes back following week and tells me the boy's taken the board to bits. Used the truckles on another board.

I'm more angry than sad.

It's the same with my surfboards and my tractor and my bike helmets.

People are always coveting the things they know I can't use no more.

Time to time Hope'll go out for the day, for work, or to cycle up some godforsaken mountain with her friends, or canoe down some freezing estuary. She'll leave me two rice cakes with butter and marmite and half a can of chicken soup to heat up on the electric hob. I'll eat the soup out of the pan with my spoon wedged into the strap I can tie to my hand with a bit of Velcro. I can get in and out of the house by pulling the strings we got attached to the door handles. I got my walking stick to pick up my hat and turn the vent on the wood burner. I need someone to come and light him, one of the carers, or Hope's mother, but apart from that I can manage on my own for a limited amount of time. I'm independent, in a limited way.

I can even empty my own catheter bag by reaching down and turning a tap with the side of my hand. Mate of mine called Toby, who was in the room with me in the hospital, invented him from an old carburettor.

Then I get a sore and I have to lie down for two months. Chair gives me a mark, and he doesn't heal, and I can't sit on him. It's a real pain in the arse. Takes the whole summer and then it's winter again. I'm hard to be around.

There are days of lightness and days of pain. They change the catheter every three weeks, but it isn't enough. I have bad nights because of the catheter followed by days of spasms. I'll get an infection, and get better, only I never gain back all the ground I lost.

I try to take care of my health. I'm always trying to cure things, religiously trying to do the things that are good for me and not the ones that do no good. I avoid cheese and cider. Co-op make lovely bread rolls, just absolute bliss to eat them, but they give me hell, so I got to stop. Food is my one enjoyment, and it gives me hell. My digestion is ruined. My stomach's collapsed because all the muscles have gone and there's no diaphragm. Nurses in the hospital called it tetra tummy, because it happens to tetraplegics. Never thought it would happen to me.

I wonder sometimes if there's any point to it and why don't I just eat what I want when I want? What does it matter?

There are problems with my tubing and they leave me weak and snappy. I try various things to relieve the pain, and none of them work.

Then one day something happens to make me reconsider my own behaviour.

Hope is out. The carers have stacked up the wood burner and left a tin of soup for me to heat up. I've had the soup and I'm dozing by the fire when the telephone rings. I've perfected a way of answering him, using the pencil to press the buttons, putting him on speaker so he can rest in my lap.

I expect it to be another one of those calls about have I had an accident that wasn't my fault. Well I always tell them, I have had an accident only he was my bloody fault.

It's not one of those calls, though.

It's Gillian. Got my number off Andy.

Cor bugger, I think, I'm going to lose it. I can't speak, and neither can she. Tears down the phone line. Both of us just crying down the phone.

After that phone call I feel sick and bad. I can feel how the left side of my heart isn't working like it should have been. It's pumping floppy and I'm thinking maybe I'm going to have another heart attack. They say premature babies are prone to heart attacks. But I'm not having another one. I say him out loud. I tell him. No way. We'll stop this time for good. We've had so much trauma.

Then Hope comes back and I can't look at her. I can't speak. I'm so ashamed of myself. I go outside to empty my leg bag in the flower beds, and I look up at the stars and they're just dots and doodles, holes cut out of the blackness.

I go back inside and Hope's washing up and I can tell she's been crying. It's the day for it. I know I ought to say something, but I can't. I hant got the words. There are some eggs and boxes displayed on the sideboard in the centre of the room. I carved them out of oak over the years and Hope kept them and brought them here.

I look at my fingers and they're all gaunt and twisted and I'm thinking shit, if only I had my hands.

I drive the chair over to the sideboard and go to try and pick up one of the boxes. I'm planning to pass him to Hope, like an apology, only he slips out of my hands and falls to the floor, and on the way down he smashes a vase that's also on the sideboard. Hope comes over with the dustpan and brush and sweeps it up, and she doesn't say a word, and neither do I.

I let her clean up my mess and I can't find a single word to say to her. I'm thinking how she's a bloody tower of strength, and how my behaviour is weakening her and weakening us, and I can't stop it.

I'm ashamed of myself and it eats away at me.

Every time I want to go outside I have to get Hope to help me put my jumper on, and my coat. We've got a routine. I'll hold out my arms and she'll get them up, one by one, and I'll lean forward and she'll drop him over my head, flatten down the back, so as not to get sores, pull down the sleeves of my long-sleeved T-shirt, so they aren't all bunched up. I can't help snapping when my arm gets stuck or she pulls it too hard.

Carers keep coming and going. No sooner will I have trained them up than they'll get pregnant and leave, or just leave. Get fed up with me and my grumpy moods.

It's Hope takes me to bed every night. She loads me into the lift, calls it my chariot, makes a picture for the back wall, all swirls and spirals of colour.

Shall I summon the chariot? she says when she catches me yawning, and it's all I can do not to weep.

One time I overheard the nurses in the hospital talking amongst themselves. If it's the male half of the couple gets hurt, nurses were saying, the woman usually stays to look after him, by and large, but if it's the woman who gets hurt, the man usually leaves.

I told Hope to go, said she should walk free, but she stayed. She stayed to care for me even after I was broke, even after we knew I was always going to be broke, that it was only going to be downhill.

I like to think I'd have stayed. I like to think I'd have stayed and spoiled her rotten, but who knows? I was a different person then.

I look at Hope and I'm reminded of how my manhood has just gone, finished, nothing, never. And that in *itself* for somebody that's been quite masculine, that's a hell of a thing to lose. Any one of those things. But there's all of them. It's hard to be cheerful.

It all comes to a head one afternoon round about teatime. I've sacked another one of my carers and Hope has found a replacement only it isn't a her, it's a he, and I says no.
  I put my foot down. I'm not having some man see me in my birthday suit and wash me and empty my bowels.
  It's bad enough women doing it, I says.

Hope's at the end of her tether but I can't be nice. I suppose I'm angry and I'm taking it all out on her because there isn't anybody else.

Things go from bad to worse between us and they carry on like that right up until the snow comes.

2013

Hope loves snow and ice. She's got a barn full of toboggans, and the kitchen's full of skates she found in the Cat's Protection, hanging up on hooks. Old-fashioned white leather lace-up boots with skates attached.

Before the accident we used to go out every year and spin around Marsh's Pool in her skates, or she'd lend her spares out to friends and they'd go off. I always worried she'd fall through the ice. We'd go tobogganing on the hill behind the cottage, come screaming down the lane and through the gateway and smash into the line of trees at the bottom of the field where the farmer keeps the rams.

When the first snow comes after the accident I'm thinking perhaps she'll give it a miss, but no. She's out the door like a horse off a starting pistol. And this happens every year. Me and Wyn sit by the window and watch her dragging her sled up the lane and then sliding back down. It's a long walk back up the lane and this time round as I'm watching I get an idea.

It takes me half an hour to get myself into my jumper and my coat and by the time I've done it I need a rest. I can't reach my hat or my gloves so I leave them off. I leave Wyn too, in case he decides to chase the toboggan. The snow is coming down thick and fast by that point, almost like a blizzard.

It's hard going at first and the chair keeps getting caught up. I have to draw on my extensive experience of driving in bad weather.

The trick is not to panic. Put him in the right gear. If you got a handbrake put him on three clicks so he'll stop the vehicle from overrunning, then take your feet off everything and let him roll along in that gear for a bit. And if you'm coming uphill you want to be in the highest gear you can, knowing your vehicle will get uphill in that gear. Just be gentle and judge it. If you see a patch of ice you RACE up to it as hard as ever you can then slowly ease the throttle and then when your wheels is on solid decent ground again you move off.

I run it all over in my mind then I gather in the wheelbarrow and we get up the hill no trouble.

Hope can't believe her eyes when she sees me emerging out of the snow. I've got Wyn's lead with me.

Tie it on, I says, then you can hold the other end and I can pull you.

I can tell she doesn't think I'm serious, she thinks the wild side of me is dead and gone. Well it's time to convince her otherwise.

Go on, I says.

We go up and down all that morning, and by the time we're finished we'm both wored out. It's the first time I seen her really smile since I fell over.

2013

There's a trailer hitched to Big Steve's car, and something inside it covered by a tarpaulin.

I'm sitting at the window looking down. Hope comes out and greets him, then she lifts the edge of the tarpaulin. I can see by her face she's excited, but I can't see what's in it. They come upstairs.

What you got there then?

Oh, he says, nothing much, just a little something.

But Hope can't help herself.

You remember those chairs we looked at? Those off-roading wheelchairs.

Cross between a go-cart and a mountain bike, I says. I remember.

Steve's got this look upon his face.

What've you gone and done?

We got you a Boma 7, he says.

You bloody din't, I says. They'm bloody twelve grand.

Well, he says, it was Lisa really. She organized it all. Concerts and cabarets, raffles. We got the choir come down, charged for entry. You name it.

I'm bloody speechless. My mouth is hanging open like a dead trout. Big Steve's wife, Lisa, doing all that just for me.

It's one or the other with people, and it isn't to do with background or nothing. I've met people with loads of money who are the meanest tightest individuals ever walked the earth, wouldn't stop for you if you were dying on the side of the road, and I've met people with no money like that too. Then there's others, like Hope's parents and Big Steve, people who'll bloody stop at nothing to help a bloody stranger.

There was a couple I used to surf with at Braunton, Christian couple, churchgoers, and they got kids. Phoned me up after the accident.

Tommy, they says, we want to help, what's your address?

What do you want it for?

We got something to send you.

Well I don't want nothing.

You can either give us the address, they says, or we'll get it from Andy.

Well they sent me £700 and they couldn't hardly afford it. I wanted to give it back, but Hope said I ought to accept it.

And now this bloody Boma 7 worth twelve bloody grand.

No, I says. I can't accept it.

Well, says Steve, I'm not taking it back. Lisa'll kick me out the bloody house, after all the trouble she's been to.

Well, I says, you're going to have to.

Well, he says, I'm not.

I look at Hope. Hope's like Big Steve. Like Lisa. She'm always helping people out on the quiet. Lending them this, and giving them that. Never tells anyone.

We can go on picnics, she says. We can go up to your piece of land.

Her face. Well I got to swallow my pride.

They winch me out of the old chair and slide me into the new one and bugger me I take off like a bloody racehorse round the bottom field, which is steep as the field where I tipped the quad bike over, and I'm thinking, I'm going to fall out, we'm never going to get back up, but we do. He's got two settings, tortoise and hare, well I can go any bloody where I want if I've got him on the hare.

I can cross the common and visit Hope's parents. I can go up top with Wyn, over the drainage channels. I can go over cattle grids. I can go up the side of a muddy hill and come down a gravel path. It changes a lot of things, as you can imagine.

I'm out in it all the time. We go for picnics, Hope on the bike and me in the Boma. We follow mountain-bike tracks. We go to the Elan valley.

We go to Aberystwyth and she brings her roller skates and we tie her to the back of the chair and go flying along the promenade. Lady sees us. That's cheating, she says.

We go to Hope's cabin where she and I got together.

Then we go up to my patch of woodland and bugger me it all comes charging back.

All that I worked for, and all that I lost.

9 May 2013

The amount of CO2 in the atmosphere exceeds 400 ppm for the first time in fifty-five years of measurement and probably more than 3 million years of Earth's history.[24]

2013

Night before we'm due to go up to the land I lie in bed thinking about all the trees I planted in my life and all the trees I cut down. I can see all the trees I cut down in one pile in my mind, and all the trees I planted in another pile in my mind, only I can't see which pile is higher and it bothers me right into my dreams.

I see them all: holly, hazel, hawthorn, cherry, apple, oak.

Ash. Elm. Birch. Pine. Scots pine, with the needles all twisted and short, pine cones looser than maritime and not in clusters, orangey bark, which is another way of distinguishing, only I can't tell the difference because I'm colour-blind, and because I'm dreaming.

The container is rusted but still intact. There isn't much left of the sawhorse, looks like a man whittled right down to his bones, everything gone from the branches except the knots, but he'm still standing upright and in position. The winch I sold. The nest boxes have fallen off the trees and they'm rotting on the ground.

Hope's had the farmer go through with his tractor and cut a path for the Boma, so we can go up without getting into difficulties. She comes up to check on the trees every month or so, but she never says much about them and I never ask. I don't know what to expect.

Well I can hardly believe my eyes.

Hope comes up behind me and puts her hands on my shoulders.
    There, she says. Look at that.

I'm cold and shivery, so we go over to her cabin on the other side of the valley and put the kettle on. She's had farmer widen the track there too, so I can get down easy, only I can't get inside the cabin because the door is too narrow and there's a step. I have to wait outside while Hope makes a pot of tea.

There's a blackbird singing and I move the chair so I've got the sun on my face and I'm looking upwards into the spreading branches of one of the old oaks and I'm listening to the blackbird and it's like the blackbird is leading me all the way back to Exmoor and I'm thinking about the way Winsford Hill smelled in August when she was covered in heather and the gorse that come up to your head and over your head.

I'm thinking how when I moved into my cottage, back in 1983,

I used to get through buckets of peanuts in the feeder. Used to get siskins and all sorts. Last time I saw Andy Harrison he told me how these days he leaves peanuts on the table and six months later he'll chuck em all away because they'll be stinking of mould.

We eat our lunch and afterwards Hope sits leaning against the big oak with the swing attached. She's got her eyes closed and her head tipped back to catch the sun. I can see the lines on her face, and some I recognize, and others are new and I wonder if I put them there myself. I want to pick her up and carry her over the threshold and make the lines on her face go away. I'd do anything for her, anything at all, that's how I feel.

But there's nothing I can do.

2013

Since the Exmoor Biodiversity Action Plan was published twelve years ago, in 2001, there have been some important changes to populations of key species on Exmoor. Some species are now known to be extinct in the park. These include iconic upland birds such as lapwing and ring ouzel, butterflies such as the pearl-bordered fritillary, and mammals such as the water vole, whilst other species including curlew, merlin, and marsh fritillary are now at very low numbers.[25]

2014

It's the day the birds changed their call. I can't hear it inside because of the double glazing, but it's very clear once I step outside. Springs are funny. Some you notice, some you don't. I notice more now. More'n I did then.

There's a woman pounding along, mud on her face and on her clothes, up to her neck in it. Never seen me. Never seen nothing. Head down, bang bang bang. Trying to wore herself out, I expect. Trying to escape her devils. Well I know something about that.

I'm parked in the lane, right in the middle of it. I've got my binoculars wedged on my arm and I'm watching a pair of kites courting.

I notice footsteps coming and I let the binoculars fall down into my lap and I see her grind to a halt. She's spotted the wheelbarrow. I can see her thinking shit, that better not happen to me, better push me off a cliff if it does.

She comes forward and explains how she's house-sitting for Hope's parents, looking after the stock while they'm off gallivanting somewhere. Says her name's Catrina. We break the ice on me being from Exmoor and her being from Cornwall, and she's into surfing, so that's common ground. I tell her how I nearly drowned at Harlyn.

It was a hurricane swell and I'd gone down with Andy Harrison as I often did. He jumped in and paddled straight out, got out there with dry hair, made it look easy, while I was all scraped up and receiving a pounding just for getting one wave wrong. I paddled for ages and ages and I kept getting hit by these monsters. Got knocked off and dragged along. Hell of a pull. Water always was the test of my sporting abilities. I'm a shit swimmer. Twenty yards and I'll drown. I'm frightened to death of water.

I'd got a new board, which was shorter than my other boards, thinner than what I was used to. I'd had him made to my own measurements, by a mate called Little Al. Once you got him on a wave he was absolutely superb, but he was a bloody horrible board to paddle, pushed a load of water up front.

I paddled for ages, and eventually I did get out there.
   Well done, Tommy mate, Andy said. Well done, boy.
   Yip all right, I said. I'm bloody knackered.
   He was still a fair way from me. I had my head down

paddling. The next thing I heard was him saying, whoa shit! Look up, Tommy, he'm coming for you.

I looked out towards the horizon and I saw him coming, big and dark, big wall of water, heading straight for me, and I wasn't far enough out, and he was coming so quick I couldn't paddle wide, so I thought, I'll paddle into him, and hope he don't break, but then he started feathering and I thought God.

I'm giving it all I can and I'm out of puff and I'm thinking, I gotta roll this one, roll my board, ketch hold of the nose, wrap my legs round and pull the front down and pray. Well I kept hold of my board and he pulled my fingernails out nearly, and then I got hit by every wave. Pushed me down until I was full of air. I'd bob up, pull my surfboard up, manage to pull in a breath, and another one'd hit me. And that happened three or four times. I'd breathed so much water I was burping every breath and I was thinking, I'm going to bloody drown here, another big one like that, I'm never going to get back. I just hant got the energy to paddle.

Well I did get back. I snowballed back to the beach, and when I stood up, the legs on my full steamer wetsuit were filled right up with water. I was like the Michelin man. I waited for Andy to catch one in and we walked back up to the car park together.

There was a chap in a chair riding towards us down the concrete path. He had a blanket over his legs, and it was all tucked in. He wasn't very old.

We stood aside to let him pass.

If I'm ever like him, I said to Andy after he was out of earshot, I want you to push me over that cliff.

Yeah all right, Andy said. I will.

Catrina's quiet.

I can see her thinking, turning it over in her mind.

Do you believe in God?

Well I wasn't expecting that.

Whose God?

I don't know. God. You said when the wave was feathering you thought God. You prayed.

Well, I says, not that sort, but yes. I believe in some intelligent person, who knows and makes everything, who is kind.

2015

An international team of eighteen researchers presents an updated report on planetary boundaries at Davos, explaining that the land-use boundary has now been breached, along with biodiversity, nitrogen/phosphorus and climate.

2015

Severe drought in the Amazon.

2016

Tis terrible precious.

I pull into a parking spot and turn off the engine.

If there's no fog you can see all the way to Dartmoor.

We're in my van, the one with the joystick instead of a steering wheel and a dashboard like a bloody aeroplane cockpit. Catrina's in the front passenger seat and Hope's in the back.

We used to come up here for picnics when I was a boy. Chicken and salt sandwiches, or dripping if it was pork for Sunday lunch. Used to pick wortleberries, which most people call bilberries, and some others call whimberries. It was me and Mother and Father and Susan. Father and Mother are up here still, scattered.

I point out the Caractacus Stone.

Old road used to go straight through, I says, only all the carts kept hitting the stone, so they moved the road.

I point out the turning to Knaplock and the field I topped when I was seventeen.

Nobody else would touch it, I says. Tractor ran away with me. Ran away four times in actual fact, and eventually I did give up. Climbed out. Walked home. Jumped out because he was tipping. Ran down the hill with tractor chasing me. Rolled five times and eventually he stopped.

There's fields quite steep all the way to Yellacombe, meadows so steep Father had to cut the bracken with a scythe. I had my own little scythe and I used to poke around with it.

I show them the hedge made out of old bottles, those ones with dimples on the bottom. They put a gurt big stone there now.

Tell her about the time you drove up a hedge, says Hope.

Well, I says, I was looking through a pair of binoculars.

I wanted to know what it was like, and I turned the corner only the corner wasn't there, so I drove up the hedge and tipped over.

I could do a whole book on your accidents with cars, says Catrina.

I expect you could, I says. Only the NHS might withdraw my privileges.

I point out Hell Manor and the road that used to lead up to my cottage. Van almost turns by itself. I wait for a gap in the countryside where I can slow up and look over. I can see the spot in the distance, on the brow of the hill, with the patch of woodland behind. I love that line of trees.

Hope leans forward again.

It was a nice cottage, she says, didn't deserve knocking down.

I don't argue, but I'm half glad they knocked it down.

Such a major part of my life's happenings took place up at that cottage, and now they'm locked away for good.

We get to the flat bit of road and I tell them about the time I reached a hundred mile an hour going along there.
Why?
Well there's a question.
I don't know, I says. Showing off, maybe, because everyone was sat out on the lawn by the pub and having their pint and I wanted to come screeching round the corner.

We pass alongside Exmoor Forest, which is not a forest in the sense of trees, but a royal forest. Hunting ground. It was owned by the king before he handed it to the Fortescues, all of it, left and right, the whole skyline, all part of the forest.

The van drops a couple of gears to go up the hill where I turned over my Bedford Beagle. I still find it strange driving an automatic, but everything is strange now. The oaks in the valley are clustered around like they'm having a meeting.

It's coming up to pheasant-shooting season and the roads are covered in dead birds. They ship them in now there aren't the wild birds to kill. Every other vehicle's a shooting vehicle plastered in mud and bearing a gamekeeper. Puts me in mind of Uncle Jim, even though he's been dead for ten years. Broke my bloody heart when he snuffed it. Went all to pieces at the funeral. Drowned every bugger else in tears. Had to walk away.

We park the van beside the tea room where Mother was borned, on an area of grass and conifers which used to be the milk stand. Farmers'd leave the churns all lined up with their names on and their numbers and the milk lorry used to come and pick them up.

It's only three bits of a mile from there to Edbrook cottages. Council houses. Ex-council houses. They been sold like everything else.

I tie Wyn to the wheelbarrow and we set off on foot past the house where Nurse Purnell lived. I used to collect her club money for my grandad on Mother's side. There was an old rickety path we used to turn into a scramble. Shortcut. The path is higher and steeper than I remember, like a present-day mountain-biking route, and I'm only in my ordinary chair.

I ask Hope to take Wyn.

Are your brakes working?

Oh aye, I was only thinking he might drag me out into the main road.

We travel slowly down over the rocky slopes and over the little packhorse bridge. There's a new plaque on one of the old houses. *Ernest Bevin, Statesman, born here on 7 March 1804.*

We trundle past Miss Dorothy's house. Miss Dorothy's parents used to let the locals have the cottages by the school and let them stay until they died and if they had no money they didn't pay rent. It was how things used to be done.

It's a big red-brick house with a garden and greenhouses. They're a bit dilapidated now and it occurs to me to wonder if Miss Dorothy is dead, but Walter Barwood's over yonder tending to Miss Dorothy's potatoes, only it isn't Walter Barwood, because I know for a fact Walter Barwood is dead.

We trundle past the house as used to belong to Sir Hugh and Lady Dobson, only the stables have been converted into cottages. Mother worked up there cleaning after Father lost his strength. I used to fish in their river and Sir Hugh used to shout at me to get out of his garden, but Lady Dobson knew I was into birds. She gave me one of my first bird books. Still got him, only all the pages have fallen out.

We trundle past the paddock where Esmond had his horses and goats and chickens, where the turtle doves had their nests, only it's houses now. Dozens of them. Little sign at the entrance: DARBY'S PATCH. At least they kept the name.

We pass the place where I left the Perazone bottle on the Corsican pine. He's gone, of course. I cut him down myself. But the oak with the hole in the trunk where robin lived is still standing. I used to know every tree by name, and I knew every bird's nest, and who lived in it. I wonder if robin's still there, or any of his descendants.

There's a wall by the side of the road. Walked alongside it every day, on my way to school, and never could see what was on the other side. It's the same now. I can't see over the top of the wall.

My chest is tight and I stop to gather my breath. I can see the roofs of Edbrook cottages. Mother's aunty lives in the end one. We live in number two. Mrs Gay lives in three. Number one is Miss Lovecombe and Mama Tunstall.

There are ducks and chickens in Father's garden.

Go up to number two, I says to Hope, ask if Mrs Carson's in.

The woman who comes out's got long hair tied back and an apron tied around her waist and plastic clogs on her feet. She's as old as I am, which knocks the breath out of me.

Hello, I says.

Tommy?

Tis me, I says.

Well, says Reenie. Cor. Bugger me. I can see her's trying not to look at my legs and for my own part I'm trying not to look at her legs, which are spilling out over her ankles like a pair of sausages. How've you been doing?

I'm all right. How about you?

Not so bad, she says. Better n'you.

Well, I says, I hant got to walk about now so I don't get tired.

Well, she says. There is that.

Tis a strange coincidence, I says, you living here.

Tis that, says Reenie. Tis that. Only we bought it.

You haven't changed, I says.

Cor, she says. Bumped into Andrew Rudd t'other day, and he don't half give you some praise. Says you taught him all he knows. Set him upon his way.

Cor, I says, well that's nice. How's Heather?

Oh, says Reenie. Didn't you hear?

Hear what?

Oh, she says. Heather's gone.

I can feel my heart pounding, the life draining out of me.

No, she says, not like that. She'm gone travelling. Last I heard she was running around the Himalayas.

Well, I says, when I got my breath back. Cor. Heather off and running around the Himalayas. It used to be terrible hard to get Heather up Winsford Hill and down again.

I don't expect she'm running exactly, says Reenie. I think she was lonely. The boys moved away. One of them went to Australia. Other one's in Birmingham I think, somewhere like that. Not a lot of life here any more. Winsford has changed some. More of a holiday place. Stops the youngsters coming in. Can't afford to buy housing. Well, such as tis.

Always was dead as a clat.

Well, no it want.

No, I says. We had fun.

Only thing we got left standing is the pub here.
Yeah.
And we don't go in that. Mr Dale is the new landlord.
Oh bloody hell.
Yeah, he's a bit of a difference.
I'll say. Them all your birds commandeered the field?
Them are.
I point to a pile of bones in the road.
What's gone on there then? Somat been hungry. I bet that's a fucken sparryhawk, brought en up here picked en clean. What d'you reckon?

One of me sparrers I expect, or a blue tit. Still, everything's got to eat.

I remember Father tipping a load of junk in the river and woods behind the house, I says.

We'm still finding bits of it, says Reenie.

Oh whoops.

Reenie looks right in my eyes, stares right in like she always used to, and I stare back and it's like staring into a hole that's been punched through time.

You should never go back. I can hear Mother's voice, sharp and clear. It's never ever the same.

Catrina's off up the lane, looking at the beech tree in the hedge, looking up, trying to find the raven's nest I expect. That's her favourite story. Hope's been over looking at the chickens and the ducks only now she's making her way back towards where we're still talking.

She'm your wife?

Not yet.

The world in front of my eyes goes dark like it does sometimes and when the light comes back Reenie's crouching down and her eyes are on my own level.

Do you need some water?

Oh, I says, no, I'm all right.

Hope stands behind the chair. We better get going, she says. Leave you in peace.

Reenie rubs her hands on her apron.

You going to call in on Miss Dorothy?

She still running?

She is, although she had a bad fall last year, spent a few months in hospital.

Well I spose we ought.

That's right, says Reenie.

We wait outside Miss Dorothy's garden until the man who isn't Walter Barwood notices us.

Is Miss Dorothy in?

Who's that? says the man, and then he says, Tommy?

Yeah, it's Tommy.

Well, he says. Cor. How're you doing?

Yeah, I'm all right, I says. We'm just down for the week. Check up on the old haunts.

Well, he says. I heard about your mother.

That's right, I says, most people heard about her. What you got there? French tatties?

Yep, she'm always liked em, only we hant had a good harvest this year. Been very wet. Wet at the wrong time.

Awful bloody weather, I says. Is Miss Dorothy in?

Yeah, says the man. She'm up in her studio. She'm had a bit of a fall since you saw her last.

I heard that.

Went to hospital for a bit of a spell but she's back painting again. Ninety-four she is, and back painting again.

Well, I says. Ninety-four.

Been a bit of a long sort of haul, says the man. Well, you'll know about that.

Long drag back, I says.

Wyn's messing around chasing his own tail.

Well, I says, we better get going.

You can go up if you like, say hello.

Can I get up there in my wheelbarrow?

Yeah, you can do. You can't get into the studio. But you can go up there.

The man opens the gate.

Don't worry about the dog, he says.

Hope goes first, with Wyn on the lead, then Catrina, and me following along behind. The man directs us up the garden path.

Which door is it?

Tis the red one, says the man, and he goes back to pulling potatoes, like he been doing when we turned up.

I hang back a bit when Hope knocks on the red door. I don't know as I've ever been inside Miss Dorothy's house before, or her garden.

Hello, comes a voice.

Miss Dorothy, I says, you got a visitor. Little Hedley boy. Arthur's boy.

I can hear the sound of the door opening and there she is, same rosy cheeks and bright blue eyes, like a bird, like robin's wife, and she looks younger than Reenie in some respects, although she's so bent over she's nearly the same height as me sat down in my wheelbarrow.

Hello, I says.

How are you?

I'm all right, I says. And yourself?

Not so bad.

Her doesn't have an Exmoor accent. Not one bit of one, even if she has lived in Winsford her whole life, borned and bred. She's known me since birth. It's a strange feeling.

Hope asks Miss Dorothy if she's still painting and she says she is still trying to do a bit only she's had to give up print-making because she isn't tall enough any more to get the right angle to see what she's doing. So she's had to stop that.

We have to move on, don't we?

We do. We have to keep going, as they say.

There's the same sign over the door at Patrick's gun shop: FISHING, GUNS, COUNTRY CLOTHING. There's a pair of horns in the window.

Look how much they're going for, I says to Hope. Two hundred and ninety-five quid for a pair of horns. You could make yourself a fortune.

There are guns in the window alongside the horns, and fishing rods and hunting stools.

Young chap comes out.

Where d'you get your skulls? I says.

Ross Campbell.

Oh, I says, is that right?

Got a whole room of them he's collected over the years.

I know that, I says. I don't mean the antlers. I mean the skulls.

Well he don't know, so I ask him if Patrick's about.

Patrick always writes to us at Christmas, even after most of the others have stopped.

Well, says the young chap. He's in the office. But it's difficult to get to.

Perhaps he could come out? says Hope.

My legs are hurting. Hope offers to go and look up in town for somewhere we can get the wheelbarrow in for a cup of tea. Not long after she's gone I hear the sound of Patrick making his way across the shop floor, dragging his bad leg with the help of a pair of crutches. He's all dressed up in a suit jacket and a white shirt and a bow tie.

Tommy, he says. Looking the same as ever.

Well I don't think so, I says.

How're you doing, mate?

I'm all right.

Well done, he says. Good.

You're not looking too bad yourself.

I'm playing in the town band tonight, says Patrick, hence the outfit. I don't normally dress up like this, but when I turned fifty I thought instead of taking tractors apart I really need to take up something I can do sitting down. So I play cornet and take tractors apart as well.

How's business? Booming, I expect.

It's not bad. It's okay. I wouldn't say he was booming.

Struggle to get better I expect, won't it? Can't fit no more shoots in, can they?

You wouldn't have thought so, would you? Only them big shoots don't come in here anyway.

Don't they really?

Bring it all with em, says Patrick. Come down here in their Range Rovers and their helicopters and they bring everything down with em and take it all back again. Only time we see em is if something breaks. Then suddenly we'm long-lost pals and it's all can you fit me in by two o'clock.

Yeah?

Well sometimes we do and sometimes we don't, and whatever we do we don't half charge em for it. They don't care.

Get a pretty penny for them horns too, I says.

That's right, says Patrick. They sell for good money. Most people who buy em, and there's a lot of females buying em for their partners and such, they got a house and they think that'll look good in the corner or up the stairs. Plenty of room, they just want a set of antlers. For most people they'm just a pair of antlers, says Patrick. But to you and me each one of them is a memory.

That's right, I says, that's the attraction of them.

You still got yours?

I kept eight pairs. Gave the rest away.

15 October 2016

The 28th Meeting of the Parties who signed up to the Montreal Protocol takes place in Kigali, Rwanda. Leaders are trying to reach an agreement to phase out HFCs, which were introduced as an alternative to ozone-depleting CFCs. It has been discovered that HFCs, now widespread in refrigerators, aerosols, foams and other products, are a major source of CO_2, with emissions growing at 8 per cent per year, and annual emissions set to rise to 19 per cent of global CO_2 emissions by 2050.[26]

2016

Mornings are the hardest part of the day. Having to go from lying down flat to sitting up.

There's people paralysed who can sit on the loo, and then there's the likes of me. I got a fat stomach, not because I'm fat, but because I got no diaphragm to keep everything in the right position. I got no muscle holding it up, nothing to keep my tummy separated from my lungs and my heart. I look fat as a pig. When I'm lying down it sags and there's nothing out the sides here. Then when I get up in the morning it's *whoa shit*, because the upper layer starts pushing down.

I broke a vertebra when I overturned the quad bike.

Hangman's fracture, surgeon called it. There's a right mess in there, he said. Didn't mince his words.

Stable fracture. If I'd done what Hope said and gone to the doctor when it happened and if he'd had me come in to the hospital and I'd lain still for long enough, I'd have healed. I'd have had a good chance of it.

But I was used to being tough, just cracking on with things, not wanting to make a fuss. I didn't go to the doctor, and I didn't go to the hospital. I pushed on through, and then my legs gave way and I fell down the stairs.

Bone split down the middle perpendicular and another one broke horizontal and a splinter came off and shot out the side and cut through my spinal cord.

2016

A group of experts advise the International Geological Congress in Cape Town that a new geological epoch, called the Anthropocene, should be declared. The experts believe the new epoch began around 1950, and expect it to be defined by the radioactive elements dispersed across the planet as a result of nuclear bomb tests. Other likely indicators include soot from power stations, plastic pollution, concrete, and even the bones left by the global proliferation of the domestic chicken.[27]

2016

It's the Sunday before Christmas, and there are carols on the radio instead of news. It's the best part of the day, when it's just me and Hope, listening to the radio, before the bloody carers come and I have to find the energy to let them haul me out of bed and start the process, let the pain and trauma start again.

I'm lying here thinking how I loved Christmas when I was a boy. How we'd get stockings with apples, brazil nuts, walnuts, socks. Lovely times we had. Used to go to Aunty Flo in Minehead and Aunty Gwen and Uncle Ron and my cousins Jill and Lorraine would come. Sometimes it would snow and we'd stay home. Stilton and snow. Two of my favourite things.

I remember the Christmas I had my heart attack. Twasn't so good. And the one when Uncle let all the stolen sheep jump out of his lorry outside our cottage and Father was mad.

Hope always gets up before I do. I got to wait for the carers.

She gets up and makes a pot of tea and gives me a cup while I'm lying down.

Who ever thought it'd be possible to drink a cup of tea while lying flat on your back?

Well Hope worked out a method.

She puts a straw in the mug and puts the other end in my mouth and I drink him like that. I can sort of turn my head and suck in the tea. There's an art to it, a technique. When Catrina does it she moves the cup and nearly bloody drowns me.

Hope gets back into her bed, a single bed, which is pushed up next to mine. She turns the radio off.

What did you do that for? I says.

Yes, she says.

Yes what? I says.

Yes I will marry you.

2017

We've reached the end of the lane, where it forks.

Which way?

You're in charge.

Let's go this way.

Yip, I said. Nice to get some sunshine. See, you go t'other way you'm in the shadow.

It's very tranquil. The sound of the chair and the sound of birdsong and the sound of the river.

What d'you think tomorra?

Weather?

Eh?

What?

What be bout tomorra?

I don't even know what that means.

What. Are. You. Doing. Tomorrow? Do you want me to speak the Queen's English?

Yes please.

What do you think you'll be doing tomorrow?

Um, I'm gunoo . . .

Gunoo? I'm a gnu?

Probably get up.

Well at some point, good idea. I'd get up myself if I could.

Catrina's back in the hedge again, pulling something out of it.

What's this?

Dog rose.

No it isn't.

Yes it is.

What's that furry thing doing in it?

I don't know. Don't pick it off. Growth, I reckon. I got one on my shoulder.

Like that?

Yep.

You've got them coming out of your ears too.

Don't you start. Chris up the road says she'm having me in next week. She going to trim my nose hair, trim my ears, trim my sideboards, do my neck.

Are you excited?

No. I'm bloody terrified.

Next morning I trundle down on my own and sit under the oak that's leaning out of the hedge by the common.

I tip the chair back so I'm looking up into the branches of the oak and into the leaves.

Trees are what we got in common, me and Hope, bind us

together. We got this tremendous love of woodlands. The first few years I knew her I was able to pass my woodland knowledge along to her, although she didn't need teaching, not really. Now it's the other way around. I learn from her. She's so intense with her woodlands, and her nature restoration, and she'll see it one way and I'll see it another way, and we'll have differences of opinion, but they'm always constructive.

The ground is thick with leaves.

We picked out an oak leaf from the piles of leaves fallen on the ground and I took it to a silversmith and he made a copy and put it on a chain. It's what she wants instead of a ring.

Back at the cottage Catrina's at the table with a cup of coffee and she's scribbling in her notebook.

Do you want tea?

No thank you.

I go to the window and swing my chair back so I'm nearly horizontal. I close my eyes. It hurts to breathe. Throat like sandpaper.

I can feel her watching me.

Are you okay?

I been bad lately, I says. Only you mustn't tell Hope. Promise me you won't.

I promise.

You tell Hope, I says, and there'll be war.

Andy Harrison turns up in the afternoon and the minute he walks in the room I know something isn't right. I'm having a rough day myself. Can't catch my breath. Body keeps going into spasms. But still I know. Andy's a peacock, type of man to stick his chest out, only now he's hunched over, and his face is grey round the edges like an old dishcloth.

Well, he says, putting two litres of cider on the table, we best get on with your stag night then.

What's that?

Never mind, he says, tis only the three of us.

Well, I says. Thank God for that.

I told him I didn't want no stag night. I didn't want nothing, and now here he is with two big bottles of bloody Strongbow.

I let him pour me an inch or two.

I got to be careful these days. Tis the leg bag. Starts to go back through the system and gets recycled in my kidneys and I get the shivers.

Catrina pours herself a gin out of Hope's cupboard and we trundle outside together. They two sit on the bench that looks out over the valley which is the last place to catch the sun. I pull up close by.

Cheers, says Andy.

Cheers, says Catrina.

Cheers, I says. Straighten my hat, would you?

She bends over to straighten my hat which is at such an angle he's practically vertical. I can see the sun on my legs and

I can almost feel the warmth. I can feel it with my mind, or perhaps it's the cider.

Birds are out in force. Parson Jack the pheasant comes hopping along, then the green woodpecker makes an appearance.

There isn't so much down there as what there is up here, says Andy.

Is that so?

I think part of it is we got a huge population of jackdaws and they just go through the nests and kill the songbirds. Magpies too.

Yep, I says, and nobody shoots em like we used to. Myself, Johnny, Horatio, we used to go around in the spring, mark the nests when the trees were bare, then go around and tip the nests out.

Andy fills up his glass again and he tries to fill mine up too, but I tell him no.

Cheers, he says. We had some fun, didn't we?

We did, mate.

Good old days, he says.

Catrina puts a steak and kidney pie in the oven, which Hope bought in the Co-op and left in the freezer, and Andy applies himself to the task of drinking four litres of bloody cider. I help him as much as I can, which isn't much, and he's the life and soul of the party, right up until he tells me he's contracted a rare form of cancer. He's had the treatment, he says, and it seems to be working, only they don't know for how long.

I try to speak only I can't. It's all gone black.

Catrina's shouting my name and Andy's got the door open and they're wafting air into my face.
    Hold my legs up, I says.
    It's happened before, like someone flicked a trip in my brain, like a power outage.
    Andy holds my legs up and we wait like that.
    What can I do? he says. I'll stay in your room if you like. Anything you need.
    No, I says. Tis nothing you can do.

2009

I'm in the hospital nine months, with every Tom, Dick and Harry coming along to poke their fingers here and there and everywhere. Modesty completely dissolved.

I've got every faith in hospitals, because whenever things have gone awry, they always put me back on the road again.
    It hasn't sunk in.

Everyone keeps telling me I'm a fighter. Get fed up with it. No, I says. I had my last fight in 1992.
    Landlord's calling time and I'm doing judo with his son, lifting him up, putting him down.

There are lots of tears, especially when people come to see me. I expect I'm not very talkative, it's all I can do to hold myself together.

Surgeon comes and stands by the bed.
 Can you fix it? I ask him.
 Words like bullets.
 No chance.

2017

I'm lying on my back waiting for the carers and running through my speech in my head.

If it's true love it bonds naturally. Love is wanting to be with that person as that person is.

I can remember one time in particular. I can't remember what we were fighting about but it was a bad fight and I took off. I got halfway down the road before I thought to myself, you arsehole, you're being an arse. I turned around and I went back, and when she opened the door I could see she'd been crying and I swore I'd never make Hope cry like that again.

There are no negative thoughts any more, from my side at least, only a deep sense of feeling between the two of us. It's been growing for years, and it's intensified over the past months. The accident put a stall on everything, but here we are, about to get married. In it for the duration.

The carers come and get me up and Hope disappears to her parents' house to put her dress on. By the time Andy turns up I'm wrecked.

You got the rings? he says.

What rings?

Wedding rings, mate.

Oh, I says, and I give him the silver necklace cast out of the oak leaf we took from the ground.

We all wait out the front of the house and it's a big old hall belonging to friends of Hope's, then Rosy the pony comes along pulling the little trap she's been trained for, and Hope is sat up there alongside her father, all decked out with flowers in her hair.

She's wearing a green dress and I look at her there in her green dress with flowers in her hair and I can't believe my luck. I can't believe any of it.

They're all standing around and crying and throwing petals and dancing and I can see how much they love her too, and I want to pull them all aside and explain how they don't know the half of it, they don't know one-hundredth of it.

Nobody cares like Hope.

I've seen it, over and over again.

The old and dying animals, the stogged and sheep-torn mountains.

Hope cares for all the broken things.

2018

It's the coldest March since records began, and Hope is up in Scotland for her work. She's supposed to be home already, but snow has fallen, snow on snow, and the roads are closed, and the trains aren't running.

Carers can't get their cars in, got to leave them at the bottom of the hill and wade through the fields with snow up to their thighs.

Then the storms pass and the sun comes out and Hope comes home. It's still cold, but not like before. I suppose I get complacent.

One day she goes out for the afternoon.

It's nine years since the accident, and I've learned to plan everything out. Everything I do. It's like going on an expedition. I'll think it all through. Weather is something I've learned to keep my eye on, and knowing distances. If I can reach that tree that's fully in leaf when it's April showers, then when it rains I can shelter underneath. If I get to the top of the track and I see rain coming down the valley I know I've got half an hour to make it back inside. Only this time I don't plan it out. I take my eye off the ball.

Hope isn't due back until six. Her mum comes over at lunchtime to stoke up the fire and give me some lunch. It's two o'clock when she leaves and I'm thinking, well, I'll take Wyn for a walk just quickly. It's cold for the time of year, but I don't try to put my coat on. It's a performance, getting my coat on over my head and then getting my arms through without someone to help them down. I'm not going to be long. I just need to give him a little spin off the lead so he'll settle and I can have my nap in peace.

Well I get as far as the bottom of the track and I notice the orchard gate is slightly open, so I go in to close it. I swivel round with the wheelchair and I succeed in closing it, only when I turn to come out I sink in the mud.

I look on the wheelchair and find a bit of rope, which I keep there for this purpose, and I try looping the rope around the gate and using the gate to get traction. I try tipping the chair back and tipping it forwards and tipping it over to one side and the other side. I try all the different ways I can think of. I exhaust my wealth of knowledge on stuck vehicles.

Then I look up the valley behind me and I can see these bloody great black clouds and I know it's going to be hailstones.

I'm thinking, Chris'll be going past at three o'clock to get the children, and generally she looks down the track. I can hang on until then.

So I hang on, and the storm comes and fills up my lap with hailstones, all down my trousers they are, and the more I try to get them out, the more they melt. Then Chris goes past to get the children, only this one time she doesn't look down the track, just looks straight ahead of her.

There's nothing I can do except wait. My other neighbour, who lives at the bottom of the track, always goes past at five o'clock, so I wait for that, only five o'clock comes and goes, and there's no sign of him neither.

I try to get Wyn to come close so I can sink my fingers in his coat and warm them up that way but he isn't interested. Bloody dog. I give him knick-knacks from my dinner plate and buy him biscuits specially, I spoil him and spoil him, and now I need him he isn't interested.

Spike would have helped me. He used to hear me coming up the garden path and stick his head through the cat flap in welcome. I got a photo somewhere, Spike with his head poking out of the cat flap. I can feel myself welling up. All the things I never think about are right there uppermost in my mind and I can't turn away from them. I'm stuck in the mud in the hail and the cold and I got no way of getting myself out.

Hope comes back half past five and she bundles me inside and stokes the fire right up and sits me in front of it and makes me drink hot tea. I close my eyes and go to sleep and I wake up and drink my tea, only he's gone cold, and I go to bed as normal and sleep the night through.

I think I've got away with it. Spend the next day by the fire mostly sleeping, then I go to bed and fall asleep again. Only a few hours later I wake up coughing and rasping like a gate that's rusted on its hinges. I can't hardly draw breath. I try to press my alarm button to wake up Hope but I can't reach it.

Well, I'm thinking, tis the end, and I remember thinking it's a lonely way to go, suffocating to death in the darkness with nobody to hold my hand. I'm worried about Hope waking up and having to find my body and how she's going to feel.

Next thing I'm in an ambulance heading up the A470 to Shrewsbury, just like before. Only this time they got bells.

Next time I wake up I can smell disinfectant, and there's lights burning into the back of my head, and the sound of the machines. It's like a bad dream, it's like one of my nightmares.

I try to open the gate and walk up through the patch of woodland and out into the area of bracken and up onto the top of the hill. I try listening for all the sounds, for blackbird and skylark and robin and owl, only I can't hear nothing except the rushing of blood in my own head, and the rapid beating of my own heart, and it's the sound of my own life swirling around, it's the sound of time, and there's nothing I can do to make it stop.

2019

I am quite literally from another age. I was born during the Holocene, the name given to the 12,000-year period of climatic stability that allowed humans to settle, farm and create civilizations. Now, in the space of one human lifetime, all that has changed. The Holocene has ended. The Garden of Eden is no more.

David Attenborough[28]

2019

Everyone used to think if you'm driving down a narrow lane with high banks there's nowhere to go, and you'm safe, but it's a false sense of security. If you stop going forwards, if you go backwards into the bank, you'll tip over. Reminds me of the way we been marching on, like time only goes one way: well it don't.

Industrial Revolution was the start of it, and it was difficult to put a limit on afterwards. Drifted away from the useful stuff to producing shit and rubbish we don't really need and all for the sake of making money and climbing up towards the top of the tree and being king of the castle.

What they don't realize is you can climb as high as you bloody like but Devil'll still be waiting down below. Farmer sent a cat across Tarr Steps to disprove it and cat disappeared in a puff of smoke.

I got so much going on with my body. I've had a pulmonary embolism. My blood is very sticky. Am I going to die of another heart attack? I don't know. Nobody knows. I know if I die of a heart attack it could be really painful. I know the feelings that lead towards it.

I could pump myself up with tablets but then it'd get so's I was taking so many tablets I'd lose sight of what I was taking them for.

Funny thing is all the way through I expected it to get more stable but he'm just getting more unpredictable, like the weather.

Tip my chair back. No. Tip it forward. Where's my jumper? The thick purple one our friend, Rachel, knitted me after the accident. There. Hook it up with the stick and put a piece of it behind me neck. Tip my chair back again. That's better.

I'm in front of the windows trying to get my thoughts in order, running the stories backwards and forwards in my mind, making sure Catrina got all the important ones.

I had the accident on Wednesday, went back to work Thursday, completed the job, loaded up the stuff, finished, wrapped up, came home.
   Friday I went to visit my friends in Newtown.
   You look really pale, go and see the doctor and whatever.
   Yeah, yeah, I'll be all right.

Made an appointment with the chiropractor for two o'clock Monday afternoon.

Saturday me and Hope gone up to the land. I remember picking Spike up at one point, chucking him over the gate, climbing the gate myself, everything fine.

Must have been bloody awful for Hope. I kept saying to her, yeah but where's me legs? Well they're here. Put me on my side, I said. Go on now. No, she said. I'm not doing it. Touch me legs, I said. Well, she said, I *am*.

Rain has turned to hail.

What time is it?

Seven thirty.

Only half an hour and Sian will be here to put me to bed. Catrina's waiting to get started on Hope's gin. She thinks I can't hear her, tiptoeing backwards and forwards to the cupboard.

Is that rain or is that hail?

I know it's hot in here – Catrina's lying on the sofa in a T-shirt and she'm taken off her socks – but I'm so cold.

Power cut'd kill me. I'd die of cold without my electric blanket and the central heating. Lift wouldn't work, nor the winch neither. They'd have to lay me out on the floorboards, in front of the wood burner, and then I'd get sores from not having my inflating mattress, and I wouldn't be able to charge up my chair.

I used to have quirky thoughts about what might happen when the polar ice caps melted. How ice is solid, and as such uneven, but when it melts it becomes liquid, and therefore even, and with that in mind, and combined with how Earth tips on her axis, how she'm tilted, does that not mean her'll level up when the ice caps have melted and turned to fluid?

No, says Catrina.

Could it upset the whole balance, change our orbit?

More likely half the world'll starve and the other half'll eat them.

Catrina likes an argument. She'm always pulling me up. Told me how birds been collecting rubber bands and feeding them to their young, thinking they'm worms.

Christ. Can't get it out of my head.

Horatio come up and he'm whingeing on about how they'm going to have windmills on the moor. Well they'm five miles up the road from him and not at all visible, but he got up this petition and they all went signing it and saying, we don't want wind power. Well, I says, you'm fast enough saying what you don't want, but what is it you do want?

2019

The Intergovernmental Science-Policy Platform on Biodiversity and Ecosystem Services reports that a quarter of all plants and animals assessed — totalling one million species worldwide — are threatened with extinction.[29]

2019

They'm all up in London dancing in the streets, them Extinction Rebellion protestors, and getting themselves arrested, and I'm not saying there isn't a need. But what annoys me with these green activists – and I'm not saying all of them – but there's some out there like to be objecting to policies and what's going on, but they'm still happy to jump on an aeroplane to go to Tenerife on a jolly.

My old man used to walk to work. Seven mile there and seven mile back, and when the weekend come he'd be working on the farms up on the moor and he'd get me to carry the tools. Axe, hook, sickle, shovel.

We were living in Winsford at the time and he was working other side of Dulverton driving a tractor. Foreman said why don't you take the tractor home? So he did. Only all the other workers got narky and reported it to the big chiefs and they didn't like it so it had to stop. That's people.

Carers said to me this morning they've given up trying to predict what I'm going to be like. It all changes so fast. Been times lately they thought this could be it.

It's only when I'm overtaken by the erosion of pain, and my mind starts working on how much pain can the body cope with before it goes pop, it's only then I get scared.

2019

40,000 fires rage through the Amazon.

2020

Storms are coming thick and fast and there isn't much anyone can do about it. We'm out of time. We've done unsurmountable amount of damage. And now there's this virus coming, and all hell could break loose if he gets up here.

It's like a bad French film, this long-drawn-out dying process. I mentioned that to Hope earlier and she did laugh. It's all I've got nowadays, the chance to make Hope laugh, and my birds, and Parson Jack, and my memories, and my ghosts.

I worry about leaving Hope on her own. When you get older you get vulnerable, and there's always someone trying to knock you off your little perch. Hope is such a strong person. But everyone breaks in the end. We'm terrible fragile, even them who think they'm strong as oxes.

Was it my strength put me in here, or my weakness?
    I still don't know.

You're a veteran now, doctor says, ain't nothing much more we can do for you.

Midwives used to suffocate babies when they weren't quite right. Stick them in the trench to make the beans grow. It was out of compassion for the babies.

But they couldn't let me die.

I'd like to choose how it happens, just go in my sleep. But I expect it'll be awful. Nothing's ever been easy in my life.

I spent all morning sleeping under that big old oak in the hedge by the common. I had the sun on my face and I was listening to a cuckoo. Used to be hundreds of them when I was a boy. They said cuckoo brings summer with him and takes it back again. Not any more.

I like just sitting there under the oak with my head tipped back listening to the sounds, the layers of sounds.

I like looking up at the leaves, all the different shades of green, only every time I close my eyes I see the Kia-Ora advert. Two crows with a hat on.

Walter Barwood had a gurt rack of a garden of his own, fair chunk of garden, and he tended for Miss Dorothy's parents. I'd sit on the wall and I'd say what sort of potatoes is that?

What sort is at en?

Pink fur apple.
Gad you're joking.
You go on home and ask your father.
He was all bent over.
George had the cottage, cottage was free, hanged hisself.

There was a woman speaking on the radio this morning about dying, and perhaps that's why I'm thinking on it so much. If you'm dying she said, you'm already got the pain. Dying doesn't make any more pain. Death is painless. It's the symptoms of what's killing you that hurt. The death rattle that everyone goes on about? That's not pain. Just proves the body is totally relaxed. The sound of the body totally relaxed.

Got friends coming to say goodbye, every bloody weekend, only they can't stay in the house, because of this virus, so they got to drive up and back in a day, and all that just to sit outside for half an hour and drink a cup of tea. I get tired after half an hour. Wored out. Got to shut up.
Catrina come up from Cornwall.
What d'you want to happen afterwards?
She'm not one to beat around the bush.
I want to go to hell, I says. What do you think?
I mean what do you want to happen to your body?
Burnt or buried?
Yep.
I'm not that bothered really. Churchyard or open countryside.

I'd like to be buried in the woods up here, next to Hope, only then I get to thinking about Exmoor, and there's a pull, you know.

She's still scribbling in her notebook.

You written this book yet?

No, she says.

Well I don't want it to go by the wayside. If I snuff it you carry on.

Okay, she says.

It shouldn't be too dark, I says, it should be funny.

2020

Worldwide, the population sizes of mammals, birds, fish, amphibians and reptiles have dropped 68 per cent since 1970.[30]

May 2021

The gate is rusty, sunk on its hinges. I have to lift it slightly to open it. I can hear the sound of the latch going and the sound of water and the sound of bees. I can hear the sound of trees rustling and swaying in the wind. I can hear aspen, her leaves rustling, and I can hear them calling.

Tommy, where are you?

I can hear Sian telling Hope, he corpsed on me, and I'm trying to come back, because I can see Hope's face. But I know it won't be long.

Dying's a wonderful gift, isn't it?

Someone has to sit with me all day now, in case I choke. I don't want to eat, because then I'll start choking, and they have to come and pump the bottom of my lungs and we'll be there for hours sometimes, and they'm on a tag team, Hope and Sian and Hope's mum and dad, and they'm exhausted, we'm all exhausted, and I think just let me die, only I'm scared of what happens next.

We've had ambulance come four times, because I can't breathe, and my oxygen levels plummet, and they can't get them back up.

June 2021

They'm staring down at me, beady eyes like a pair of bloody sparrers.

Where were you?

I went somewhere, didn't I?

You did. We lost you.

I don't think I'm getting up today. I croak it out. I think twould be physically impossible.

It's hard to talk.

Well that was Friday. Monday I'm worse again. Sian can't do the normal care because I'm choking and spluttering and my oxygen levels are going down.

Let's get him comfortable.

Well that doesn't work. They're moving me this way and that way, pressing buttons, getting the bed to tilt up and down and back like a bloody fairground ride, and eventually they've got me in a kind of V-shape and that's the best we can manage.

Sian comes in on Tuesday.
   You don't work here no more, I tell her.
   I think you'll find I bloody do, she says.

Tuesday they get me sedated, only when I come round the terror's got me too, and Sian's sitting there and I'm thinking, what happens if I can't breathe and I can't die, then what?
   It isn't the pain so much as the fear of choking.
   If you'm ready to go you can go, says Sian.
   Help me.
   Even just those two words cost more'n I got.
   I can't, she says, and I can see she'm crying.

Wednesday there's a hell of a storm. The oak in the hedge by the common comes down and blocks the track.

Thursday they're all clustered around. I ask them to leave and I pull Hope in close.
   Fuel lines in the chainsaws need cleaning, I tell her.
   You haven't got any tractors, she says. You sold them, remember?
   I try again, and it ain't no easier the second time.
   Fuel lines in the chainsaws need cleaning.
   Oh, she says, okay.
   And Owen Gray is a man who needs reminding.
   Owen Gray brings the logs and I don't want her getting cold.

I'm holding onto Hope and I don't want to let her go only my hand is slipping. I got no strength left. I got nothing left.

2021

Over the past few decades, scientists from many disciplines have joined forces to put together the most comprehensive picture of what the hell just happened.

Initially, they considered the start of the Industrial Revolution as the moment when humanity became a planetary force. But that view has changed recently. There is now overwhelming evidence that it was really the 1950s when humanity began to overpower Earth's life-support system.

Johan Rockström[31]

16 June 2021

Hope is leaning down and kissing me on the top of the head and I'm running down the lane and there's elm and ash in the hedgerows and I'm inside the hedge and I'm climbing the beech and I'm looking up and it's all new leaves and very green and I can hear Father calling.

Git down out of that tree.

I'm not coming down. You'll beat me.

And Father's walking off down the lane in the moonlight and I'm looking after him and my heart's beating so hard he'm going to push himself out of my chest and I can't breathe and Father's coming back and he's cutting the tree down and I'm hanging on for dear life and then I'm falling, falling, down, down, through all these other trees and all these shades of green and I'm sort of like flying through the air and I'm thinking shit, I'm going to hit the ground, only I don't.

Before the tree ever touches the ground Father's up the other end and he's got me. He's ketched me. He's ketched me right out of the sky.

PART THREE
DEEP TIME

See then that ye walk circumspectly, not as fools but as wise, Redeeming the time, because the days are evil.

Ephesians 5:15

Three Weeks Later

Hope picks out a place in the woods near the cabin, a resting place. There's space for her, too, for when her time comes.

The coffin is made of timber from the oak in the hedge that came down. Drew and Tyson bring an Alaskan sawmill and cut him up there in the lane. They make a bench with what's left.

Hope goes to Rhodri's farm and picks a rock to mark the spot. Time's been washing over him for 450 million years or thereabouts. They use a tractor to drag him out of the river where he has lately fallen.

There's flowers on the coffin, sweet peas and sweet william from people's gardens, and wild things from the hedgerows.

There's family and friends from here and from there, and a couple of strangers.

There's music, and sloe gin, and rain, but not much.

They fill in the hole with shovels Hope brought down, a spadeful each. They each say their bit.

And night falls and more nights, nights after nights, and they go back to their lives and jobs and tasks and houses, and the coffin and its contents remain on the side of the hill, tangling with the roots of oak and hazel and birch and bramble, tangling with the bones of robin and blackbird and sparrow and owl, and the stones and the fungi and the bacteria, and time washes over, and breaks us into smaller and smaller pieces, into particles . . .

Five Months Later

In my head and mind I don't know how to stay hopeful when the world is collapsing before my eyes at this accelerating pace. I even often feel guilty, for not being able to stay hopeful all the time, beat myself up for not being able to take more action, for spending most of my energy trying to stay sane.

Then a voice in me nudges, come on, Ecem, you know despair won't help creating the world you want to live in. It's going to be a long fight, but we can do hard things, show up even when we're messy and afraid.

We need to sit in the grief, we have to be willing to acknowledge the pain, the painful truth.

Ecem Albayrak, speaking at COP26 in Glasgow[32]

10,000 Years Later

The fossilized bones of 60 billion chickens are found on every continent on Earth.

240,000 Years Later

Half of the unstable nuclei in a sample of the plutonium that leaked out of Chernobyl's nuclear reactor have decayed.

10 Million Years Later

Earth's biodiversity has recovered to the level it was before the mass extinction that happened during the Anthropocene.

4.5 Billion Years Later

Half of the unstable nuclei in a sample of the atomic nuclei of U-238 – the depleted uranium used in the nuclear weapons testing that began in 1945 – have decayed.

. . . and we swirl forth as clouds of dust and gas and the small particles of swirling dust and gas draw together, and are bound by gravity into larger particles, and solar winds sweep away the lighter elements, like hydrogen and helium, leaving only heavy rocks, and small worlds are formed out of these spinning rocks.

In memory of Hedley Ralph Collard
1953–2021

You ask me where I'm going
You want to know where I've been
You ask me how I got here
Well I've forgotten most of what I've seen

I could tell you of cabbages and kings
Love and the treasure love brings
And I could tell you of utter loneliness
And how that's when the blackbird sings

You want to hear all of my stories
How can a life go this way?
Well I'll tell you I lost my balance
One ordinary day

But the trees they keep on growing
And the stars don't know that they died
And there's a hundred million light years
Still flickering in my eyes.[33]

Acknowledgements

Thanks to Ralph's family especially, Carol and Keith; and his friends especially Andy H., John P., Steve and Lisa, Kate, Willow, Patrick, Miles, Chris up the hill, Rachel, Andy and Tuppin, and others, for being part of this story.

Thanks to all of Ralph's carers, and special thanks to Rhian, for answering my difficult questions about the end of his life.

Thanks to John Akomfrah, whose installation Purple was a huge inspiration.

Thanks to Johan Rockstrom and Owen Gaffney for their book, Breaking Boundaries (and subsequent Netflix documentary of the same name) which helped me find a context and structure for this story.

Thanks to Jay Griffiths, for introducing me to Ralph and to Llanidloes, and for being a steady source of comfort, inspiration and encouragement, both with your friendship and your work.

Thanks to my agent, Jessica Woollard, for your dogged determination to help change the world through books.

Thanks to everyone at riverrun, especially my editors Jon

Riley and Jasmine Palmer, for your patience and dedication, and publicist Joe Christie, for your skill and enthusiasm.

Thanks to Charlie Worthington and Jane Irvine, for providing me with the space and silence to write.

Thanks to the Society of Authors and the Royal Literary Fund for the very generous financial support, without which this book would not have been possible.

Thanks to my family and friends in Cornwall, for always being there.

Thanks to climate scientists and activists across the world for fighting so hard to try to save us from ourselves.

Thanks to Hannah Scrase, for your generosity, warmth and friendship. Knowing you has enriched my life.

Thanks to Andy McNamara, for being the light and water that allows me to keep on growing.

Notes

1. 1907–64. American marine biologist, author and conservationist, whose influential book *Silent Spring* (1962) was met with fierce opposition by chemical companies, before spurring a reversal in national pesticide policy and a nationwide ban on DDT and other pesticides.
2. Boomerang-shaped region of the Middle East that was home to some of the earliest human civilizations.
3. Rockström et al., 'A Safe Operating Space for Humanity', *Nature* magazine, nature.com, 29/9/2009.
4. WWF Living Planet Report, 2018.
5. e360.yale.edu
6. www.culturalsurvival.org
7. https://essd.copernicus.org/preprints/essd-2019-152/essd-2019-152.pdf
8. *Exmoor's Wildlife*, 1979.
9. b. 1936. American ecofeminist philosopher and historian of science most famous for her theory (and book of the same title) *The Death Of Nature* (1980), in which she identifies the Scientific Revolution of the seventeenth century as the period when science began to atomize, objectify, and dissect

nature, foretelling its eventual conception as composed of inert atomic particles.

10   nationalgeographic.com
11   Patricia Villarrubia-Gómez, PhD candidate and research assistant at the Stockholm Resilience Centre; stockholmresiliencecentre.org
12   world-nuclear.org
13   1941–2001. American environmental scientist, educator and writer. Best known as the lead author of *The Limits to Growth*, 1972.
14   In 1992 the team released an updated report, called *Beyond The Limits*. Already in the 1990s there was compelling evidence that humanity was moving deeper into unsustainable territory. *Beyond The Limits* argues that in many ways we have already overshot our limits, expanded our demands on the planet's resources and sunk beyond what can be sustained over time. (www.donellameadows.org)
15   Johan Rockström and Owen Gaffney, *Breaking Boundaries* (Penguin, 2021), p. 79.
16   Corryn Wetzel, *Smithsonian* magazine, 17/9/2021.
17   William F. Laurance, Philip M. Fearnside, Susan G. Laurance, Patricia Delamonica, Thomas E. Lovejoy, Judy M. Rankin-de Merona, Jeffrey Q. Chambers, Claude Gascon, 'Relationship between soils and Amazon forest biomass: a landscape-scale study', *Forest Ecology and Management* 118 (1–3): 127–138 (14 June 1999).
18   Jason Hall-Spencer, saveourseasmagazine.com, June 2016.
19   *Limits to Growth: The 30-Year Update*, Routledge, 2004.
20   Jason Hall-Spencer, saveourseasmagazine.com, June 2016.

21 Director of the Global Systems Institute and Chair in Climate Change and Earth System Science at the University of Exeter. His reading of Jim Lovelock's books on Gaia when he was an undergraduate ignited his passion for studying the Earth as a whole system, forming the foundations of his research to date.
22 Timothy Lenton, Hermann Held, Elmar Kriegler, Hans Joachim Schellnhuber, 'Tipping Elements in the Earth's Climate System', pnas.org, 12/2/2008.
23 e360.yale.edu
24 natgeo.org
25 Exmoor Wildlife Research and Monitoring Framework, 2014–20.
26 unep.org
27 To define a new geological epoch, a signal must be found that occurs globally and will be incorporated into deposits in the future geological record. For example, the extinction of the dinosaurs 66 million years ago at the end of the Cretaceous epoch is defined by a 'golden spike' in sediments around the world of the metal iridium, which was dispersed from the meteorite that collided with Earth to end the dinosaur age. (Damian Carrington, 'The Anthropocene Epoch: Scientists Declare Dawn of Human-Influenced Age', *Guardian*, 29 August 2016.)
28 English broadcaster, natural historian and author. He is best known for writing and presenting, in conjunction with the BBC Natural History Unit, the nine natural history documentary series forming the *Life* collection, a comprehensive survey of animal and plant life on Earth.

Speaking at World Economic Forum, Davos, 2019. (*Breaking Boundaries*, p. 47.)

29   UN Report: Nature's Dangerous Decline 'Unprecedented'; Species Extinction Rates 'Accelerating'; un.org

30   Living Planet Report, 2018.

31   b. 1965. Swedish professor and joint director of the Potsdam Institute for Climate Impact Research (PIK) in Germany. Executive director of the Stockholm Environment Institute from 2004–12. Joint author, with Owen Gaffney, of *Breaking Boundaries*, Penguin, 2021. (*Breaking Boundaries*, p. 59.)

32   Climate activist living in Istanbul. Member of the Youth Advisory Council at the Human Impacts Institute; humanimpactsinstitute.org. Student of Political Science and International Relations at Turkish-German University. Speaking at COP26 in Glasgow, November 2021.

33   Download at https://catrina-davies-songs.bandcamp.com/track/cabbages-and-kings